THE AQUATIC APE HYPOTHESIS

Elaine Morgan

'Why are there human beings?' or 'What caused them to evolve?' must be two of the most fundamental questions in science, and indeed form the whole *raison d'être* for anthropology.

Robert Foley

SOUVENIR PRESS

This paperback edition first published 1999

First published in Great Britain in 1997 by
Souvenir Press
an imprint of Profile Books Ltd
29 Cloth Fair
London
EC1A 7JQ
www.souvenirpress.co.uk

Printed and bound in Great Britain by
Clays Ltd, Elcograf S.p.A.

The moral right of the author has been asserted.

A CIP catalogue record for this book is available from the British Library.

ISBN 978 0 28564 361 1
eISBN 978 0 28563 981 2

Contents

List of Illustrations

TABLES

Acknowledgements

I owe gratitude to more people than I can remember for information, references, answers to questions, and general advice over the years. Hundreds of communications came from general readers who sent letters of appreciation, queries, ideas, and cuttings. I would like to thank them all.

Among the scientists who supplied comments, information, advice, or permission to use their material, I would like to thank the following (inclusion in this list does not imply any degree of agreement with the aquatic hypothesis): Leslie Aiello, Sir David Attenborough, Michael Chance, Bruce Charlton, Michael Crawford, Stephen Cunnane, Richard Dawkins, Frans de Waal, Christopher Dean, Daniel Dennett, Derek Denton, Robin Dunbar, Derek Ellis, Peter Rhys Evans, Karl-Erich Fichtelius, Robert Foley, John Gribbin, David Haig, Kevin Hunt, Chris Knight, Robert Martin, Desmond Morris, Michel Odent, Caroline Pond, Vernon Reynolds, Graham Richards, P. S. Rodman and H. M. McHenry, Erika Schagatay, Phillip V. Tobias, Marc Verhaegen, Peter Wheeler, Tim White.

I appreciated the opportunity of debating AAT for several months on the Internet. My thanks are due to those who supported me, and to those of my opponents who were constructive in their criticism and generous in sharing their specialist knowledge.

My thanks also to: Amanda Williams, who drew the pachyderms and the proboscis and patas monkeys; Frans de Waal, for the bonobo photographs; Jessica Johnson and

Michel Odent for the water baby photograph; Dr Terence Meaden and June Peel for the drawing of the Venus of Willendorf; Brooks Krikler Research for picture research.

Preface

This book, like the others I have written, is addressed primarily to the general reader, so I have tried to use plain and accessible language.

This time, however, I have included numbered references. After disputing for 25 years with professional scientists, I have learnt to respect the high standards they set themselves, and expect from others, in identifying their sources.

The references are unorthodox in one respect: a small percentage does not relate to the written word. The study of natural history owes a considerable and growing debt to the camera crews who travel the world recording the behaviour of rare species in inaccessible places. It is time that this material was accorded, as source material, equal status with the observations of a scientist with a notebook.

The idea on which this book is based does not qualify as a theory in the strict Popperian sense adopted by scientific philosophers—it is more accurately a 'hypothesis'. But the acronym AAT (Aquatic Ape Theory) has been in use for so long that it would be confusing to change it now.

I hope that the book will be as enjoyable to read as it was to write.

1

Death of a Hypothesis

We are back to square one.
Phillip Tobias[1]

The question

It is generally agreed that around eight or nine million years ago there lived in the forests of Africa an animal known to anthropologists as the last common ancestor (l.c.a). The descendants of the l.c.a. split into different lineages, and their extant survivors are gorillas, chimpanzees, bonobos and humans. Of these, humans differ more markedly from the African apes than the apes differ from one another. There are numerous and striking physical differences, and at least some of them began to appear either at the time when the human lineage diverged from that of the other apes, or very shortly afterwards. It would seem reasonable to conclude that something must have happened to our ancestors which did not happen to the ancestors of the other apes.

The question at issue is simply: WHAT HAPPENED?

Twenty years ago, that was regarded by anthropologists as a pertinent question, and most of them were convinced that they knew the answer to it. Today that confidence has so far evaporated that some of them query whether an answer to it is possible or even desirable. In 1996, the title of a public seminar in London was 'Do we need a theory of human evolution?'[2]

13

The savannah-based model

The explanation current in the 1960s was very clear and straightforward. The divergence between apes and humans was said to be due to climatic changes which resulted in the dwindling of the African forests and the rapid expansion of a grassland eco-system—the African savannah.

The apes, it was concluded, are descended from the populations of the last common ancestor that remained in the trees, whereas humans are descended from populations which were driven out of the shrinking forests and forced to make a living on the savannah. A minority version claimed that they were not victims but pioneers, who opted to move out to a more exciting and potentially rewarding life on the plains. In both versions the change of venue was held to account for the adaptations specific to the hominid line, such as bipedalism and nakedness and, in the longer run, to tool-making, increased intelligence and verbal communication.

It has been repeatedly asserted (for example, on the Internet) that there was never such a thing as the 'savannah theory', that it was simply a straw man constructed by Elaine Morgan for the pleasure of knocking it down again, and that no reputable scientist can be shown ever to have used the phrase 'savannah theory'. The last part of the statement is perfectly true. I would no more have expected them to use that phrase than I would expect a Creationist to refer to 'the God theory'—their faith in it was too strong for that.

But the savannah-based model was no straw man. Raymond Dart, who discovered the first African hominid fossil (the Taung baby), on what is now the veldt, believed that the harsh conditions on the savannah turned the first hominids into hunters and killers, and that this was the driving force that made us human.[3]

The savannah scenario was embraced from the beginning with enthusiasm. Lyrical popularisers like Robert Ardrey exulted in it: 'We accepted hazards and opportunities and

necessities of an order quite different from life in the trees
... More and more we lived beneath the open sky, less and
less beneath the forest canopy'.[4]

Clues to the origins of the human social order were
increasingly sought in the behaviour of savannah species
like the baboon, as researched by Irven de Vore.[5] Scientists
like John Pfeiffer (in the days before Sarich and Wilson)
commented that 'Hominids, or members of the family of
man, have spent 25 million years foraging on the savannah
and only a few thousand living in cities'.[6] Fossil hunters
like Richard Leakey recommended that 'we should take a
look at the long extinct ape that eventually ventured from
the forest fringe to live in the open'.[7] Television series like
Jacob Bronowski's *The Ascent of Man* summed it up:
'Human evolution began when the African climate changed
to drought: the lakes shrank, the forest thinned out to savannah'.[8] Numerous scientific papers seeking to explain
bipedalism or the loss of body hair saw them as natural—
even predictable—consequences of the move to the grasslands. Many people have found it hard to relinquish this
model. In Peter Wheeler's widely quoted series of papers
the savannah continued to be identified, as late as the 1990s,
as the environment where bipedalism would have proved
adaptive: 'The thermoregulatory advantages conferred by
bipedalism to a large-brained primate on the African
savannah...'[9] 'The results indicate that the equatorial
African savannah would have been a difficult habitat for
hominids to exploit'.[10]

In 1987, looking back on the earlier years, Randall L.
Susman described it as follows:

The themes of territoriality, meat-eating and regi-
mented social organisation for the exigencies of life on
the savanna recur in *all of the theories of human origins*
[my italics] that have been proposed in the 30 years
following Dart's announcement of Australopithecus
(Dart, 1926 *et seq.*; Washburn, 1957; Oakley, 1961a).[11]

That was no straw man. It was a seminal, near-universally accepted scientific paradigm.

The doubts creep in

With Donald Johanson's discovery of the fossil AL-2881 ('Lucy') at Hadar in 1974, the savannah hypothesis began to fall apart. Until then bipedalism had been generally believed to have followed after the move to the savannah, and to be one of its earliest consequences.

But Lucy did not fit easily into that picture. For one thing, she lived too early. Her bones were dated at 3.5 million years ago, but she already showed unmistakable signs of being at least partially bipedal.

She did not die in a savannah habitat, but in a wooded and well watered area of Ethiopia; and some scientists argued that there were features of her anatomy—such as the curved phalanges of her fingers—which indicated that she still spent part of her time in the trees.

A note of caution began to creep into references to the savannah model. Elizabeth Vrba found evidence of a dramatic increase in the number and variety of the herds of grazing animals in Africa at around 2.5 million years ago.[12] That opened the possibility that the emergence of savannah conditions might have been the cause of the emergence of a more advanced type of hominid, *Homo erectus*, which appeared at around that time. But it certainly contradicted the idea that savannah conditions precipitated the split between apes and hominids.

Life on the plains as the driving force of human evolution—once treated as fact—was by the 1980s written of in the subjunctive, as a possibility only: 'The emergence of hominids from earlier apes may have been associated with a shift in the environment from life in the wet forests to drier grassland and savannah. The necessity for rapid movement in the open may have provided the selective advantage for the development of bipedality'.[13]

Or then again it may not. Gradually opinion has hardened round the latter possibility. When Richard Leakey re-wrote his book *Origins* in 1992 in the new version entitled *Origins Reconsidered*, he brought the scenario up to date. 'In fact, the great plains and the immense herds on them are relatively recent aspects of the African environment, much more recent than the origin of the human family.'[14]

Two years later, in 1994, a paper in *Science* affirmed of the Tugen Hills fossil site in Kenya: 'Open grasslands at no time dominated this portion of the rift valley.'[15]

A headline in the *New York Times* commented on the increasing dissension: 'Fog thickens on climate role in the Origin of Humans'.[16]

After the savannah model

The savannah-based model can now safely be considered defunct. Bernard Wood wrote in *Nature*: 'The savannah "hypothesis" of human origins, in which the cooling system begat the savannah and the savannah begat humanity, is now discredited'.[17]

K. D. Hunt wrote in *The Journal of Human Evolution*, 'Recent evidence suggests that the common supposition that australopithecines were grassland adapted is incorrect'.[18]

And Phillip Tobias observed in a lecture in London: 'All the former savannah supporters (including myself) must now swallow our earlier words in the light of the new results from the early hominid deposits ... Of course, if savannah is eliminated as a primary cause, or selective advantage of bipedalism, then we are back to square one and have to try to find consensus on some other primary cause'.[19]

If this constituted a scientific revolution it has been on the whole a velvet one. By imperceptible stages references to the habitat of our earliest ancestors began to be modified. They changed from 'savannah' to 'savannah mosaic' and then simply to 'mosaic', to denote a patchwork environment where tropical forests and woodlands were interspersed with tracts of open country. Many of the younger anthropologists are unaware that any change has taken place. They remember hearing references to open spaces, mosaics and diversified landscapes throughout the ten or so years they have been studying the subject, and the 'straw man' charge seems to them a credible one. And even if savannah *was* the wrong word—so what? Now we have the mosaic theory instead.

Unfortunately, we don't. We have a greatly watered-down version of the savannah theory. 'Mosaic' in this sense is the description of a type of environment. It does not specify a new hypothesis, clinging as it does to the shreds of the old one. Factors like carrying behaviour and sentinel behaviour and thermoregulation are still regularly canvassed. Formerly they were attributed to the rigours of spending 100 per cent of a lifetime on the savannah. Now they are often attributed to the rigours of occasionally crossing the open spaces between one patch of forest and the next.

Yves Coppens in 1994 published a persuasive hypothesis called the 'East Side Story', suggesting that the formation of the Rift Valley some eight million years ago split the population of the Miocene apes into a group to the west in the humid forest, and a group to the east in a somewhat drier environment.[20] The East Siders were supposed to be our ancestors. The following year doubt was thrown on this idea when a French team led by Michel Brunet discovered an *afarensis* jawbone over a thousand miles to the west of the Rift Valley. But even if it had been true, it would not have solved the problem of why, at the north end of the Rift Valley, Lucy had begun walking on two legs a million years before the East Side turned into savannah.

The original savannah model—though it did not stand the test of time—was argued in strong and clear terms. We are different from the apes, it stated, because they lived in the forest and our ancestors lived on the plains. The new watered-down version suggests that we are different from the apes because their ancestors, perhaps, lived in a different part of the mosaic. Say what you will, it does not have the same ring to it.

Perhaps it is natural to hang on to the last shreds of the savannah paradigm. Apparently such tenacity is not a rare phenomenon. Robin Dunbar refers to one philosopher who defended the practice: 'Lakatos also made an important practical point when he observed there is no point in discarding a framework theory just because there is evidence against it. Without a framework theory, we cannot ask questions or design experiments ... It is much better to carry on using the old discredited theory until such time as the alternative appears.'[21]

To date, the only genuinely alternative hypothesis is the aquatic one.

This is the idea that during the fossil gap—before five million years ago—the ancestors of the hominids passed through a stage of a semi-aquatic existence before returning to a predominantly terrestrial lifestyle. It was mooted by Max Westenhofer in 1942[22]—but ignored and forgotten; then it was independently arrived at and aired by Alister Hardy.[23] As far as I can trace, only one scientist at that time (the zoologist, H. B. N. Hynes) publicly acclaimed it as 'very convincing'.[24] Years later, Colin Groves described the general reaction:

> When Sir Alister Hardy in 1960 put forth his hypothesis that, as one newspaper headline put it, 'Dip in Sea Turned Ape into Man', there was an initial stirring of discomfort from evolutionary anthropologists, then silence.
>
> Only Elaine Morgan's support of the hypothesis has kept it alive all these years ...[25]

Most professional anthropologists remain extremely wary of it, although their specific reasons for rejection have not been widely publicised. Daniel Dennett, in his book *Darwin's Dangerous Idea*, commented:

> During the last few years, when I have found myself in the company of distinguished biologists, evolutionary theorists, paleoanthropologists and other experts, I have often asked them just to tell me, please, exactly why Elaine Morgan must be wrong about the aquatic theory. I haven't yet had a reply worth mentioning, aside from those who admit, with a twinkle in their eyes, that they have also wondered the same thing.[26]

There may be sound reasons why the aquatic model, like the savannah one, will in the end after careful scrutiny have to be abandoned. But there is no case for rejecting it out of hand. Over the past ten years it has been adjusted and modified to meet valid objections and to accommodate new data. For those who have assumed that there is something inherently untenable about it, it is time to think again.

2

Where the Hominids Died

Let the fossils speak for themselves.
Palaeontologists' maxim

Palaeontologists

The Aquatic Ape hypothesis suggests that the events which diverted our own ancestors along an unusual evolutionary path had something to do with water. The evidence for this belief is derived largely from comparing human anatomy with that of apes, and will be examined in detail in the succeeding chapters.

However, if the evidence for comparative anatomy was in conflict with evidence from the fossil remains, it would be worthless.

In all conferences and discussions about human origins it is the palaeontologists whose opinions command the most respect. Only fossils provide the final conclusive evidence that those ancient ancestors—not quite apes and not quite humans—actually walked the earth and are not merely figments of a Darwinist's imagination. Moreover, it is hard won evidence, recovered at the cost of months of laborious preparation and fund-raising, followed by more months of sweat and toil and patience under a sweltering sun and with no guarantee of a successful outcome.

Their work also demands an intimate knowledge of skeletal anatomy so that they will know what they are looking for and recognise it when they find it. If Raymond Dart had not been an expert neuroanatomist, the Taung child's skull that came into his hands in 1924 would have meant

21

nothing to him.[1] As it was, he recognised the signs of its incipient humanity and continued to believe in them throughout the 22 years when everybody of any consequence assured him that it was the skull of a chimpanzee.

So it is generally accepted that it is only the fossil hunters who can tell us where the Australopithecines lived, and when they lived, and what they looked like. But that statement needs qualifying in two respects. Lucy's bones cannot quite tell us what she looked like. They can tell us whether she was tall or short, but not whether she was fat or thin, smooth or hairy. And they cannot tell us with certainty where she lived. They can only tell us where she died. As all palaeontologists are vividly aware, it is not quite the same thing.

Where the hominids died

In the Rift Valley sites almost all the hominid fossils that have ever been found—from the oldest to the most recent, from Tanzania in the south to Ethiopia in the north—are the remains of creatures who died by the water's edge.

There is, of course, nothing surprising about that. It is equally true of other mammals, simply because water-borne sediments provide the ideal conditions for fossilisation. Terrestrial mammals like antelopes and baboons may have been living on the dry savannah, but no traces would remain of the ones that died there. Their bones would be scavenged by hyenas, trampled and dispersed by migrating herds, and any remaining fragments baked brittle by the sun and reduced to dust. Those of which we have records are the small minority whose bones happened to sink into water or silt or mud and thus be preserved.

This skew in the record is known as 'taphonomic bias' (Greek *taphos* = tomb). It is one of the reasons why the savannah scenario was able to survive for so long, because one of the first lessons taught to fossil hunters is not to be fooled by taphonomic bias. In attempting to reconstruct the lifestyle of an extinct animal the water factor should be treated as irrelevant. Most of them lean over backwards to obey this injunction.

One classic example of this can be found (or rather, cannot be found) in Donald Johanson's book *Lucy*.[2] It was a popular book, packed with every fact about the discovery of that famous fossil which he thought the general reader might find interesting. But he failed to mention a detail that at least one of his readers would have found interesting—the fact that Lucy's bones were found eroding from sand which also contained the remains of crocodile eggs and of turtle eggs and crab claws. It was not because he wanted to conceal the fact; it was because he had been conditioned not to register it as a fact of any significance. In another connection he clearly describes this state of mind:

> There is no such thing as a total lack of bias. I have it; everybody has it. The fossil hunter in the field has it. If he is interested in hippo teeth, that is what he is going to find, and that will bias his collection because he will walk right up to the other fossils without noticing them.

In his paper to *Nature*[3] he went into more minute detail, but even there the crocodiles, turtles and crabs were only referred to obliquely, as a touchstone for how perfectly the contents of that stratum had been preserved. There he was addressing his peers, and did not feel the need to specify these items. They would assume from their own experience that such things might be present, and share his conviction that they should be disregarded. Later, there were speculations about whether *Australopithecus afarensis* fed on leaves or fruit or seeds. No one mentioned the possibility that Lucy's last supper might have consisted of turtle eggs and a couple of small crabs.

It could reasonably be claimed that the savannah hypothesis of human evolution was never based on the fossil record. It was based on a speculative extrapolation of the fossil record, 'corrected' to allow for taphonomic bias. It was known that hominid remains had been found in association with a wide variety of other creatures—ancient pigs and elephants and baboons and giraffes, and crocodiles and turtles and water snails and catfish. Logic alone would suggest two possible explanations. Either the hominids, like the baboons, had come to the lake or river to drink and unfortunately expired there, or else, as with the hippopotamus, it was their chosen habitat. But only one of these possibilities was considered worthy of consideration.

Richard G. Klein described the nature of the fossil evidence with exquisite precision:

> The site concentrations and specific site locations reflect the occurrence of sedimentary traps with good conditions for bone preservation, and both the australopithecines and early *Homo* surely ranged far more widely within tropical and subtropical Africa, into areas where fossils have not been found or may not exist.[4]

'Surely' in this connection is an idiomatic English usage meaning 'We are not quite certain, but any reasonable person will agree it is a pretty safe bet'. It is a safe bet that at some point or other they did begin to range widely, but the fossils cannot tell us when that began to happen—five million years ago, or four or three, or perhaps only two million years ago.

A lot also depends on the nature of the sedimentary traps. The phrase could be taken to mean dwindling water holes in the middle of a wasteland of grass and scrub. On the other hand, it could refer to a river bank or a marine delta or a lake shore. The coastline of Lake Victoria constitutes a sedimentary trap extending in excess of 2,000 miles. These inland waterways may not have been where the hominids first originated, but they would have supplied ample opportunity for their descendants, such as *Homo habilis* and *Homo erectus*, to enjoy them as a favoured kind of habitat to which they may have become accustomed.

Wet and dry places

One of the tables in Robert Foley's book *Another Unique Species* records the various habitats where hominid fossils have been found.[5] It seems to indicate that most of the sedimentary traps were more than water holes. It has a number of references to savannah and grassland, but refer-

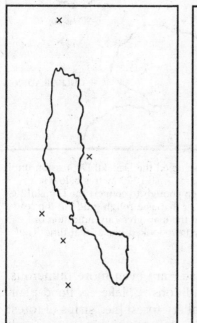 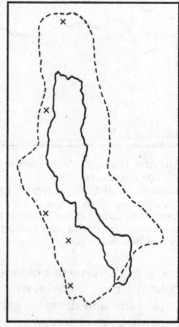

The Turkana basin is a fertile source of hominid fossils. The map on the left marks the sites of famous discoveries—clockwise from bottom left Kanapoi, Logatham, Turkwel, Nariokotome, Omo and Allia Bay. They are all now arid savannah sites. The dotted line in the right-hand map indicates the area occupied by the lake 4 million years ago. (After Meave Leakey, *National Geographic*, **vol. 188**, no. 3.)

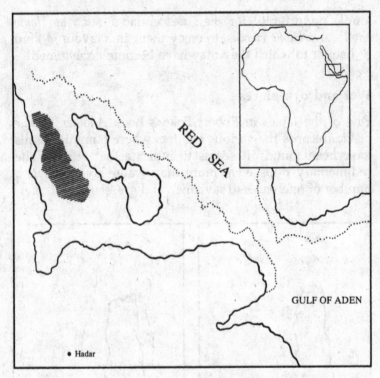

The Afar Triangle. The Red Sea occupied the Danakil Depression until the late Pleistocene. The hatched area is still 120 metres below sea level. At one time the sea rose high enough to convert the Danakil Alps into an off-shore island. Using that as a rough guide to the sea level at the height of the flooding, the map gives an indication of what the shore line of Africa may have looked like at that time. The dotted line shows the existing coastline.

ences to the presence of water are even more numerous: 'lake margin ... fluvial conditions ... lake ... flood-plain ... streams and rivers ... gallery forest [i.e., strips of forest fringing lakes or rivers] ... river banks ... swamp ... salt lake ... fresh rivers ... small lake ... flood-plain'. It is all a far cry from the standard text book illustration of a prognathous biped gazing out across a vast empty landscape relieved only by one or two stunted thorn bushes.

There is plenty of other evidence that the Rift Valley in

Table 1. Palaeoenvironmental reconstructions of early hominid localities in Africa (from Robert Foley, *Another Unique Species*).

Locality	Palaeoenvironment
Tabarin	Lake margin, with locally variable savannah elements (Hill, Drake et al. 1985)
Middle Awash	Fluvial conditions with extensive tectonic activity associated with the formation of the East African Rift Valley (Kalb et al. 1982; Clark & Kurashina 1979; Clark et al. 1984)
Laetoli	Grassland savannah with well-defined wet and dry seasons (Hay 1981)
Hadar	Lake and associated flood-plain, with braided streams and rivers (Aronson & Taieb 1981; Gray 1980)
Omo (Shungura)	After 2.1 Ma ago dry thorn savannah flanking river banks with gallery forest and swamps; before this date the environment was probably more forested (Bonnefille 1976, 1979; Brown 1981)
Koobi Fora	Before about 1.6 Ma ago, a fresh lake with flood-plains, gallery forest and dry-thorn savannah; during later times the lake fluctuated from fresh to brackish (Isaac 1984)
Olduvai	Salt lake with surrounding flood-plains with seasonal streams and rivers and dry grassland savannah; tectonic changes after 1.5 Ma ago resulted in the drying up of the lake (Hay 1976, 1981: Isaac 1984)
Peninj	Open grassland surrounding a salt lake, fed by fresh rivers (Isaac 1984)
Chesowanja	A small lake with surrounding flood-plain and grassland (Bishop, Hill & Pickford 1978: Isaac 1984)
Transvaal	Makapansgat, Sterkfontein, Swartkrans, Kromdraai, and Taung: all these were mosaic savannah environments, with Makapansgat Member 3 and Sterkfontein Member 4 being less open (more bush/woodland) than Swartkrans Mb1 and Sterkfontein Member 5 – this suggests a trend from wetter to drier conditions through time (Vrba 1975, 1976, 1985)

the past contained more water than it does now. Meave Leakey commented about the Turkana site in the *National Geographic* magazine:[6] 'For much of that time a lake far larger than today's Turkana filled most of the basin, yielding crocodile, fish, and turtle fossils but few terrestrial animals'. At Hadar, too, the lake periodically expanded to fill the whole basin,[7] and the Danakil Depression to the east of it was at one time part of the Indian Ocean.[8] Later it became land-locked, forming an inland sea which slowly evaporated. Today the central area is a vast landscape of solid salt, below sea level, thousands of feet deep and extending to the horizon in all directions. It is not possible to determine exactly how much of Ethiopia was flooded by the sea, but geologists are able to detect the high-water mark reached when the Danakil Alps were an off-shore island.

In Foley's table, there are two headings where water is not mentioned. One is not part of the Rift Valley: it is the Transvaal in South Africa, where the Taung baby was discovered. These South African sites are all caves, and it is not easy to deduce where the creatures in them lived and died, because their bones were moved after their death.

Some were probably carried into the caves by predators. Others may have been washed down a shaft into a subterranean cavern together with silt and other debris in a spate of surface flood-water. Even the Transvaal, it is now believed, was wetter then that it is today.

The other atypical site is Laetoli. It is grassland now and was grassland when the hominids lived there. This—plus the fact that it was the site of the famous footprints—meant that it was the jewel in the crown to savannah theorists. Nowhere else did they find the remains of hominids that had died on the open plain. What is different about Laetoli is that its sediments are not water-borne but air-borne.

Laetoli lay in the shadow of the highly active Sadiman volcano. There was always the odd chance that a body, or part of one, might be covered (like the inhabitants of Pompeii) with a fall of volcanic ash before it could be eaten or disturbed. Something like that happened to the footprints.[9] There had been a fall of ash followed by a shower of rain, which made a mixture rather like wet cement, perfectly designed to receive the imprint of the feet of two bipedal travellers who walked across it. Within a short time another fallout covered them up and preserved them. This is proof positive, is it not, that the savannah was the natural habitat of our early ancestors in their lifetime?

That is a possibility but not a certainty. The volcanic dust also preserved the prints of other creatures which were crossing the same stretch of land at the same time, and they were not all savannah dwellers. They have been described as an atypical 'mixed assemblage',[10] which would not normally have been found together anywhere in such numbers and such variety. They included arboreal monkeys and one galago—usually confined to a habitat high in the forest canopy. The possibility has to be considered that they were refugees, driven out of their normal habitats by fear of the volcano and possibly by fire. This transient population may well have included the makers of the bipedal footprints.

AAT model and some objections

In short, the fossil record is perfectly compatible with the supposition that at some time between eight and six million years ago, at the north end of the Rift Valley where the most ancient hominid remains have been found, one section of the l.c.a. population found itself living in a watery environment and—whether by choice or under duress—began to adapt to a semi-aquatic existence.

According to this model, some of their descendants in the succeeding millennia migrated southwards along the chain of rivers and lakes and waterways of the Rift Valley where their bones are found today. Once they had become partially adapted to, and physically dependent on, plentiful supplies of water they would have been reluctant to move too far away from it. Their days as big game hunters roaming far and wide across the arid savannah would have had to wait—firstly, until the savannah eco-system became established, and secondly, until they learned to fashion containers to carry water with them.

Among the objections that have been repeatedly raised to this scenario, one is about crocodiles, and the more substantial one concerns the skeletal physiology of the fossils themselves.

The crocodile argument is that Lucy and the First Family of *A. afarensis* could not have lived by or in water because they would all have been eaten up by crocodiles and become extinct. It is something on a par with saying that they could not have lived on the grasslands because they would all have been eaten up by lions.

In any case, the Afar crocodiles that bask in the River Awash are not man-eaters; they are too lazy and too well fed on the plentiful supply of freshwater fish.

> The muddy river is alive with crocodiles, but is also well stocked with their staple diet, catfish, and the ample supply of food may explain why the Afar tribesmen fording the river with their camels and goats cross without apparent fear or even watchfulness.[11]

This description by Colin Willock is endorsed by Donald Johanson:

> Some members of the expedition would not swim at all because of the infestation of the river by crocodiles. But these were smaller than the man-eating monsters of Kenya and Uganda, and did not seem to be consuming any of the Afar people, who were in and out of the water constantly. After a couple of weeks most of the scientists were bathing daily.[12]

As for the conformation of the skeleton of the earliest hominids, it has been claimed that if they were or had been passing through an aquatic phase their fossil remains would show unmistakable signs of it, as do the skeletons of seals and dolphins; their legs would have become shorter instead of growing longer and their arms would have begun to turn into flippers.

This argument shows a total misconception of the course of events that AAT proposes.

Whales and dolphins have been aquatic for about 70 million years and seals for between 25 and 30 million years.[13] For most of those periods the cetaceans have been fully aquatic, never returning to land; and the seals need to go ashore only to breed. The hypothetical aquatic phase of the ancestral apes during the fossil gap would have been brief, a matter of two or three million years. Nobody has suggested that they turned into mermen and mermaids. They would have been water-adapted apes in the same sense that an otter is a water-adapted mustelid. If we knew nothing of the otter except what we can deduce from its bare bones, it would take a clever scientist to detect that it was any more aquatic than its cousins the stoats and the polecats.

In the more orthodox scenarios there is constant disagreement among the experts about the meaning of the bones in every new specimen that is unearthed. To take the example of curved phalanges in the fingers of the earlier hominids: some curvature would perhaps be predictable, confirming that they were descended from arboreal creatures. The degree to which they are curved in *A. afarensis* may indicate that Lucy still spent some of her time in the trees. On the other hand, it could mean that the process of straightening them out was a slow one, not driven by any particular selective pressure, and was still incomplete. The current consensus is that *A. afarensis* did spend some time in the trees, but her skeleton indicates that when at ground level she walked on two legs. What her skeleton cannot tell us is whether at ground level she was walking on dry land or wading through water.

Regarding the legs, it is over-simplistic to say that an aquatic environment inevitably shortens the hind limbs. In cranes and herons it has dramatically lengthened them. Besides, the great increase in relative length of the hominid's hind limbs only occurred after it had been perfecting its bipedal stride on land for millions of years. It certainly did not pre-date Lucy. The proportions of her limbs are virtually identical with those of the bonobo.[14]

What part of the mosaic?

The problem with the mosaic as the backdrop of the ape/human split is that if Africa was covered with alternating patches of forest and woodland and grassland, it is harder to conceive of apes and protohominids being subject to different selective pressures. Any intrepid member of the l.c.a. who ventured out onto the stretches of grassland would be strongly motivated to return to the shade of the trees at midday to avoid the noonday heat, and again at night to climb into the branches and sleep out of the reach of predators. In that case they would have continued to interbreed with the more conservative members of the troop, and no new species would have arisen.

However, it is not reasonable to suppose that the forests and the grasslands were evenly disposed in a kind of chequerboard pattern. Trees need plenty of water. The forested patches would have been concentrated possibly on high ground where there was more rainfall, and certainly in areas where there was more ground water—in the vicinity of rivers and lakes.

A typical extant area of Africa which could be described as mosaic is that surrounding Lake Victoria. It is less like a chequerboard than a series of concentric circles. Nearest the lake is forest; further out more open woodland; then grassland; and finally desert.

Almost all the speculations on human origins share a strong preconceived idea of which of these concentric circles our ancestors inhabited. Occasionally, they state it as a matter of fact: 'In the most marginal habitat of all evolved the ancestor of the Hominidae'[15]—the margin here denoting the 'sparsely wooded, relatively arid regions'.

There is no reason for making this assertion other than habit, long ingrained during the decades when the savannah theory was king. It would be no less arbitrary, but certainly no *more* arbitrary, to affirm: 'In the wettest habitat of all evolved the ancestor of the Hominidae'.

3

Before the Biped

Why, of all the mammals that have ever walked the earth, did only one group choose to walk erect?

Donald Johanson[1]

Why bipedalism is rare

We now know that habitual bipedalism is one of the basic hallmarks distinguishing the hominid lineage from that of the African apes. It may have been the first of these hallmarks to emerge. It was certainly one of the first.

Science was slow to recognise its primacy. Earlier in the century one school of thought, led by Elliot Smith, believed that the ancestral ape first evolved a large brain, then became bipedal while still living in the trees, and only finally came to earth. W. K. Gregory believed that bipedality did not emerge until after the hominids had acquired technology and greater social complexity. In books for the general reader it was not presented as a problem. Once the ancestral apes had moved onto the savannah and become killers, it seemed the most natural thing in the world that they 'became more upright—fast, better runners'.[2]

The most dramatic sign of the volte-face in this respect took place in 1964. Until then, to qualify for the designation of *Homo*, a hominid had to exceed a certain cranial capacity. At the very lowest estimate it had to possess a brain volume of 700 cc. Some authorities thought it should be as high as 800 cc. This requirement was known as the 'cerebral rubicon'. No stipulations were laid down about the modes of locomotion. In theory a sufficiently large-brained quadruped would still have qualified as Man. It was only with the discovery of *Homo habilis* that the definition was recognised as unworkable. Louis Leakey, Phillip Tobias and John Napier wrote to *Nature* with the proposition that 'if we are to include the new material in the genus *Homo* (rather than set up a distinct genus for it which we believe to be unwise) it becomes necessary to revise the diagnosis of the genus.'[3] They suggested that the cerebral rubicon should be reduced to 600 cc and be accompanied by other hallmarks of humanity—a precision grip in the hand, and 'habitual bipedal posture and gait'. It was the way we walked, rather than the way we thought, which first set us apart from our anthropoid cousins.

Donald Johanson's question at the head of this chapter is therefore a crucial one. We have had a great many suggested answers to the more usual form of query: 'Why do humans walk on two legs?' Most of them stress the alleged advantages we have derived from bipedalism. Few of them tackle the problem of why, if it was such a thundering good idea, it was not adopted more widely by other mammalian species.

It was not as if the hominids invented bipedalism, or pioneered it. There was a time when it was one of the commonest ways of covering the ground. Bipedalism was frequently the chosen mode of locomotion among the dinosaurs although they were descended from quadrupedal

Bipedalism was not always a rare form of locomotion.

ancestors. Quite a number of them remained bipedal even after attaining colossal dimensions and becoming the masters of the terrestrial ecosphere. Creatures like *Hadrosaurus* and *Tyrannosaurus* walked everywhere on two legs—left right, left right—while their front limbs dwindled to such feeble and useless appendages that they seemed on the way to disappearing altogether like a kiwi's wings. Then they became extinct, and bipedalism went out of fashion as a method of walking the earth until the hominids arrived.

The probable explanation is that quadrupedalism in reptiles is not particularly effective. They move with four limbs typically projecting outwards from the sides of the body, and bending downwards at the knee and elbow as in crocodiles. The centre of gravity is not in the same axis as the feet, so the reptile has to throw its body into a slight horizontal curve at every step forward, and must exert considerable muscular effort to raise itself clear of the ground on its widely spaced feet.

As a mode of locomotion this is not too hard to improve on, especially if you want to move fast. In order to achieve a burst of speed, some modern lizards, like the collared lizard of the United States and the frilled lizard of Australia, rise up and run on their hind legs only, eliminating the need for the rocking motion that otherwise wastes energy and slows them down. Bipedalism in dinosaurs may have arisen in a similar fashion.

Another disadvantage of reptilian quadrupedalism is that it is not very good in the trees. Because of their splay-footed gait, it is impossible for a reptile to run freely along a branch that is no wider than its own body, and only the smallest ones can live in trees. So far as we know, there was never an arboreal dinosaur.

The forest was therefore the safest place for the little mammals that lived in the age of the dinosaurs. They could scuttle through the undergrowth or run along the branches, and it may have been the exigencies of branch running that remoulded the disposition of their limbs, so that all four legs were repositioned directly underneath the body and supporting it.

This new improved form of terrestrialism was transmitted to their descendants, which include all the major groups of terrestrial mammals. It has been perfected over something like 70 million years, and it works so well that it has scarcely ever been abandoned by any species of terrestrial mammal. It is near-universal because, other things being equal, it is unbeatable.

There are one or two exceptions. (In the living world there are one or two exceptions to almost every rule.) A few species of small desert rodents, for example, such as the jumping mouse and the kangaroo rat, have opted to hop on two legs rather than walk on four. Anyone who has tried to run over soft, dry, yielding sand will recognise that in those conditions it is impossible to work up any speed, and perhaps in those conditions strong hind legs and a leaping gait give these creatures a better chance of escaping predators.

Marsupials, too, have their own problems. In Australia, the kangaroos occupy the same ecological niche as the grazing herds of placental mammals on other continents. However, the kangaroo continues to carry her young until it is almost too big to climb into the pouch, and its long legs can be seen sticking out at the top. If the kangaroo had persisted with quadrupedalism and tried to put on speed by racing away like a springbok, her offspring would have been in peril of being jolted out every time her front legs hit the ground. Under these conditions it pays to be perpendicular.

But, apart from a handful with special problems, all the animals which have abandoned quadrupedalism have done it because they have abandoned walking on the ground. They have become air-borne (bats) or aquatic (dolphins) or aboreal.

Arboreal mammals

This last group is the one to which our own ancestors belonged. Small arboreal primates continue to be quadrupedal in the same way as squirrels do. But some larger monkeys and apes have become too heavy to run along the tops of branches and different species have dealt with the problem in different ways.

The larger lemurs in Madagascar have solved it by the strategy of leap and cling. They evolved very long, powerful legs, enabling them to launch themselves into space from one tree and cover quite long distances before landing on and clinging to the trunk of another. Some large monkeys and small apes have specialised in brachiating— travelling suspended beneath the branches, swinging from hand to hand. The adult orang-utan proceeds more cautiously, supporting its weight as far as possible with a minimum of three hand-holds at any one time.

What about the last common ancestor? This question was debated for many years. Some features common to apes and hominids suggested that they were all descended from a primate which was adapted for brachiation, while other features suggested that they could never have been true brachiators like the gibbon. Terms like 'semi-brachiator' and 'modified brachiator' were suggested. In 1975 Russell Tuttle was already pointing out that this problem had 'nagged the anthropological community for three-quarters of a century'.[4] By 1981 a near-consensus had been arrived at, which envisaged the last common ancestor as a generalised climber which was in the habit of holding on to overhead branches with one hand. Leslie Aiello summed it up:

> Recent research in comparative anatomy and locomotion of extant primates suggests that the below-branch adaptation and, particularly, a fore-limb-assisted suspensory climbing adaptation during feeding would have provided the necessary preadaptations for the

morphology observed in extant apes and human beings.[5]

All these strategies have to some extent disabled the animals which resort to them from achieving normal efficient quadrupedalism when they find themselves at ground level. Quadrupedal locomotion is an unbeatable system for an animal with four limbs of roughly equal length. But it is difficult or impossible for an animal with disproportionately long legs—like the leap-and-cling lemurs or, for that matter, *Homo sapiens*—and also for a species with disproportionately long arms like the gibbon.

On the rare occasions when these creatures come to earth they are forced to become bipedal. The leap-and-cling sifaka travels in ungainly 2-legged hops, like a child in a sack race. The ultra brachiators like the gibbon and the spider monkey cover short distances by running on two legs, holding out their very long arms to the sides to help them balance, like a tight-rope walker's pole. They display no skeletal adaptation to bipedalism in the pelvis or anywhere else, because these excursions are so brief and infrequent.

The gibbon theory

These are our closest bipedal relations. They have been described in some detail because one of the more reputable bipedalism theories is that we may have become bipedal for the same reason as the gibbon. This idea is known as the hylobatian (gibbon-like) hypothesis and has recently become even more reputable because—unlike many of the alternatives—the case for it was not weakened by the discarding of the savannah scenario. It makes it possible to minimise the whole problem of human bipedalism by implying that it was predictable—a short step for an ape and no big deal for a hominid—'especially,' H. M. McHenry commented, 'given the fact that all the lesser apes today are habitual bipeds'.[6]

There is a sense in which that statement is true, but there is less in it than meets the eye. 'All the lesser apes' here refers to the Asian gibbons and siamangs. They are bipeds only to the extent that on the infrequent occasions when they engage in a non-suspensory form of locomotion they proceed on their hind legs. The dynamics of the gibbon's mode of bipedalism bear little resemblance to ours: '. . . its locomotion is unique among the primates'.[7] Strictly speaking, they are not habitual primates but habitual brachiators, as John Napier described them: '[The] dominant component in their locomotion is arm-swinging'.[8]

The crucial question about the gibbon theory is whether or not, at the time of the ape/hominid split, the African l.c.a. had arms so disproportionately long that it was virtually obliged to be bipedal on the ground. There is no evidence that this was the case.

The commonest assumption is that the last common ancestor was a generalised climber rather like a chimpanzee. In that case, the hylobatian theory postulates that subsequent to the ape/human split the hominids' ancestors changed into gibbon-like brachiators, came to the ground in that condition and had to walk bipedally, and later their limb proportions reverted to the ratio we find in australopithecines. To date, however, there is no evidence that this ever happened.

The great apes

Humans are not the only descendants of the l.c.a. to have moved down to live at ground level. The gorillas have done the same. Although they still sleep in the trees, very little of their waking activity takes place above ground. It has been calculated at about 2.9 per cent on average, and less than that in the case of adult males.[9] Savannah chimpanzees also spend a lot of time away from the trees during the wet season. It might be expected that gorillas, chimpanzees and humans, sharing as they do a common arboreal ancestor, would, on coming to the ground, face precisely the same locomotion problems and solve them in precisely the same way.

That did not happen. Chimpanzees and gorillas opted to walk on four legs, and hominids on two. Something was different. Either they had become different while still in the trees, or when they came to ground level it was under different environmental conditions.

Once an arboreal mammal has begun to adapt to brachiation or any other suspensory activity, it is not easy for it to re-learn quadrupedal walking, even when the relative length of its front and hind limbs does not rule it out altogether. All the great apes can and do walk on four legs, but they are not very good at it.

The difficulty lies in their hands. The hands of gorillas and chimpanzees have become specialised for branch-holding. They evolved a hooked grip for larger branches and a double-locking grip for slender ones. This resulted in a limited degree of backward extension (dorsiflexion) both at the wrist joint and in the fingers. As Frederick Wood Jones pointed out, 'Every Anthropoid Ape perforce flexes its digits and walks upon its knuckles, for it is a well-known anatomical fact that an Anthropoid Ape is unable to extend the joints of its fingers, its wrist, and its elbow simultaneously'.[10]

Thus, in order to walk on four legs, chimpanzees and gorillas are constrained to walk on their knuckles, and have evolved leathery pads on the weight-bearing joints of the second to fifth fingers. The orang-utan is even more specialised for branch-hanging and has to walk on its fists.

Knuckle-walking and fist-walking are comparatively inefficient ways of covering the ground, far more expensive in energy terms than that of the average quadruped. The world's most committed hanger-on, the sloth, has limbs so specialised for suspension that it cannot walk at all, and has to proceed at ground level by clutching at any surrounding vegetation and dragging itself along.

It is not generally believed that our own ancestor ever passed through a knuckle-walking stage.[11] The old idea that we are descended from ancestors who walked ape-fashion on four legs in the forest and only straightened up when they reached the savannah has been abandoned, for several reasons. The only human quadrupeds—our children—have no difficulty in crawling with their hands flat on the ground like a baboon. And while apes and chimpanzees have been knuckle-walkers for so long that their babies are born with knuckle-pads, our own show no traces—even prenatally—of ever having passed through that stage.

Not only did they not go through a knuckle-walking stage. It appears that they did not, subsequent to the split, pass through a stage of quadrupedalism. H. M. McHenry is emphatic on this point. 'Could these and other australopithecine traits of the forelimb imply a terrestrial locomotor behaviour other than bipedalism? With available evidence from the hindlimb the answer is an unequivocable no. All diagnostic hindlimb specimens of *Australopithecus* show a complete reorganisation for bipedalism, and forelimbs show no sign that they were used as a quadrupedal prop.'[12]

Yet there was nothing in their morphology to preclude them from quadrupedalism when they descended to earth. Their arms and legs were near enough to the same size to make it perfectly feasible for them. Even as late as Lucy, although 'a modest degree of hindlimb joint enlargement had already taken place . . . the highly-derived relative joint size characteristics of modern humans had not yet been achieved'.[13] Other things being equal, Lucy's index of relative limb length would have enabled her to walk on four limbs as readily as the bonobo which shares the same index.

But the option was not taken up by the ancestors of the hominids. From the moment they descended there is no evidence that they ever put their hands to the ground, either with fingers flat or knuckles bent. They simply climbed down and stood up to walk.

Why?

There is—as there has been throughout most of the century—a wide variety of possible answers. But in the aftermath of the velvet revolution, how many of them will still stand up?

4

Walking in the Mosaic

*Theories on the origin of bipedalism itself are by their very nature impor-
tant and speculative.*

H. M. McHenry[1]

'Four legs good—two legs better', proclaimed the headline
of an article in *Nature*.[2]

It is natural for any member of a species which has been
perfecting bipedal locomotion for five or six million years
to perceive it as an improvement on anything that has gone
before. Yet there have always been scientists who, trying
to imagine the first stages of the transition to walking on
two legs, have found the whole concept bizarre.

John Hunter earlier in the century wrote: 'The erect position of man is probably the worst calculated for either
natural offence or defence of any natural posture. His body
becomes wholly exposed; it is even unfit for resisting the
force of either wind or water'.[3] John Pfeiffer pointed out,
'Animals balancing themselves on two feet are easier to
knock over, more conspicuous, less agile when it comes to
dodging and feinting and escape tactics'.[4]

C. O. Lovejoy thought that for any quadruped to get up
on its hind legs was 'an insane thing to do',[5] and noted
that 'It deprives us of speed and agility and all but elimin-
ates our capacity to climb trees which yield many important
primate foods, such as fruit and nuts'.[6] Stephen Jay Gould
found it the hardest of all human features to explain.
'Upright posture is the difficult event, the rapid and fundamental reconstruction of our anatomy'.[7]

John Napier observed that 'Human walking is a unique

44

activity during which the body, step by step, teeters on the edge of catastrophe'.[8] F. Wood Jones could not conceive how a creature in such a posture could have survived without weapons 'unless his environment was in every way peculiar'.[9] And doctors are well aware that perpendicular walking has proved a costly option for which we are still paying today by increased liability to a wide range of disorders, from lower back pains to hernias and haemorrhoids.[10]

On the other hand, there has always been, and still is, a school of thought which holds that erect bipedalism emerged because it was the most effective mode of locomotion open to our ancestors when they first embarked on it. Since we have become aware that bipedalism preceded the advent of savannah conditions and the use of tools and weapons, this interpretation becomes particularly appealing. If it was simply the most effective way of getting from A to B, then it could have evolved anywhere—on the forest floor, or in the clearings, or on the woodland/grassland interface—as easily as on the savannah.

There are two criteria for measuring its relative efficiency. One is speed, and the other is energy costs. If our predecessors found they could move faster on two legs, or if moving on two legs consumed fewer calories (or preferably both), then no further explanation would be needed. They sound simple calculations to make, but the argument about them goes on and on.

The prima facie case against the greater efficiency of bipedalism looks pretty devastating. To begin with, running on two legs is slower than running on four. During each step taken by a quadruped, four limbs have contacted the ground and actively contributed to its forward motion—whereas in our case it is only half as many. The other two limbs may be energetically thrusting backwards and forwards, but they are not contributing to our progress by pushing the air aside, as a bat's wing does: they are just working hard to keep us upright, by maintaining our centre of gravity in the right position. Stability is a much trickier problem for a vertical mammal than it is for a horizontal one.

Secondly, running on two legs is energetically very expensive. Compared with average quadrupeds of the same size, we expend roughly twice as much energy as they do in running the same distance. David R. Carrier discussed this 'energetic paradox' in 1984.

> The energetic cost of transport (oxygen consumption per unit body mass per unit distance travelled) for running humans is relatively high in comparison with that for other mammals and running birds. Early comparative studies showed that a mammal the size of man should consume roughly 0.10 ml. of oxygen per gram body mass per kilometre travelled (Taylor, Schmidt-Nielsen and Raab, 1970) but the measured value for man is over twice this amount (0.212).[11]

Not unnaturally, many people felt there was a problem about a creature which voluntarily opted for a unique mode of locomotion which was simultaneously slower and energetically more costly than that of other mammals. The search was on for the missing factor which would make sense of the data.

It was difficult to represent Lucy's rather clumsy bipedalism as the consequence of anything, since it preceded all the other datable changes, both physiological and ecological. It was a great temptation to see it instead as a harbinger of something—as a preparation, as a pre-adaptation, a halfway stage to something wonderful: to the free-striding, weapon-toting *Homo* who roamed the grassy plains a couple of million years later.

Roger Lewin in 1987 described this kind of thinking as 'the teleological trap', the idea that features could begin to evolve because they would—when perfected—be adaptive at some future time in conditions that had not yet arisen.

The first apes to practise habitual bipedal walking were envisaged, however subconsciously, as having an end in view, a goal that would not be achieved in their lifetime.

Lewin included in the list of those who fell into this teleological trap 'virtually every paleoanthropologist who has put pen to paper on the subject of human origins'.[12]

Energy costs of bipedalism

Since 1970 one of the goals of anthropologists has been to erase the image of human bipedalism as recklessly profligate of energy.

The obvious first step was to stop comparing the human condition to that of an average quadruped, or that conceptual creature 'a mammal the size of a man'. That was not comparing like with like. It was out of the question that any descendant of the l.c.a. would be able to cover the ground as efficiently as animals like horses or greyhounds or cheetahs. The hominids' ancestors had traded in running speed for the ability to clamber around in, and swing from, the branches. At ground level they were inevitably out of their element.

So the proper procedure was to compare them with other primates. If it could be proved that human bipedal locomotion was more effortless than the gait on the ground of other primates, then it was reasoned that its emergence would require no other explanation. It would be enough to say that the hominids discovered the most efficient way for an ex-arboreal to cover the ground, and that was on two legs.

(It would not answer Johanson's question: Why didn't the other apes also discover it? But there is a school of evolutionists which considers that 'why-not-questions' are less legitimate than 'Why-questions': they do not feel called upon to address them.)

The first—and, in many ways, still the best—piece of research into the question was conducted by C. R. Taylor and V. J. Rowntree, who published their results in 1973. They posed a straightforward question, stated in the title of their paper, 'Running on two or four legs: Which consumes more energy?' and devised a way of answering it.[13]

They trained two chimpanzees and two capuchin monkeys to run on a treadmill, either on two legs or on four legs as required, and measured the amount of oxygen consumed at various speeds. There was no danger of not comparing like with like: in each case they were comparing energy costs of quadrupedalism and bipedalism *in the same individual.*

They did not state in the paper what result they had expected, but they did admit that the outcome surprised them. The energy consumed, regardless of the speed of locomotion, was exactly the same when the animals ran on two legs as when they ran on four. Taylor and Rowntree were not into hypothesising, and advanced no theories about what might have been the reason for resorting to bipedalism, but they were pretty clear about what was not the reason:

> Thus the cost or efficiency of bipedal versus quadrupedal locomotion probably should not be used in arguments weighing the relative advantages and disadvantages that bipedal locomotion conferred on man.

That might have been expected to be the end of it. There was peace for about seven years, but it did not last.

In 1980 the whole question was opened up again in a paper by P. S. Rodman and H. M. McHenry.[14] They made some valid points.

One was that locomotion does not necessarily mean running. For much of the time—probably most of the time—it means walking. And in the case of humans that makes a great deal of difference. In our case, running on two legs is 75 per cent less efficient than walking on two legs.[15]

Rodman and McHenry therefore concentrated on walk-

ing speeds. And they further stressed that human locomotor efficiency should not be compared with that of the
average mammal, or even the average primate. In the light
of our evolutionary history, the relevant comparison must
be specifically with other descendants of the l.c.a. They
chose the chimpanzee, and stressed that the chimpanzee—
still an actively arboreal creature—is at ground level one
of the most inefficient walkers in the whole of the animal
kingdom. The chimpanzee consumes 50 per cent more
energy when it walks than a true quadruped of similar
size. Consequently, when they compared the energy costs
of walking in quadrupedal chimpanzees with bipedal
humans, Rodman and McHenry proved conclusively that
human bipedalism is *'considerably more efficient than quadrupedalism of living hominoids'* [their italics].

For anyone who leans to the opinion that bipedalism was
an inevitable development rather than an astonishing one,
this conclusion has been welcomed as the final solution to
a long-standing problem. The paper itself made this claim:
'We conclude that bipedalism bestowed an energetic

Table 2. Comparative energetic costs of walking for quadrupedal
chimpanzee and bipedal human at normal travel speeds[14]

Speed	Subject	Body weight	Predicted cost[1] (mlO_2/g/km)	observed cost (mlO_2/g/km)	Observed/ predicted cost (× 100)
2.9 km/hr[2]	Chimpanzee	17.5 kg	0.351	0.522	149%
	Human	70.0 kg	0.225	0.193[3]	86%
4.5 km/hr[4]	Chimpanzee	17.5 kg	0.287	0.426[5]	148%
	Human	70.0 kg	0.180	0.170[3]	94%

[1] Cost for a true quadruped of the same weight, predicted from equation 4 of Taylor et al.
('70): $M'_{run} = 8.5W^{-0.40} + \frac{6.0}{V}W^{-0.025}$; W = weight (g); V = speed (km/hr).

[2] Average speed of male chimpanzees in the wild.

[3] Value estimated from the fitted relationship of E_m, the energy expenditure per metre
walked, to walking speed of Zarrugh et al. ('74): $E_m = \frac{32}{V} + 0.0050 V$; V = speed (m/min),
units converted to mlO_2/g/km. Similar results are given by Margaria et al. ('63).

[4] Normal, and optimal, human walking speed (Ralston, '76; Zarrugh et al., '74).

[5] Value estimated from the fitted relationship of oxygen consumption to velocity for
quadrupedal chimpanzees of Taylor and Rowntree ('73): $M'_{run} = 0.25 + \frac{0.79}{V}$; V = speed of
walking (km/hr).

advantage on the Miocene hominid ancestors of the Hominidae', and McHenry repeated it two years later in terms which would be relevant in the context of the new mosaic backdrop: bipedalism '. . . could have arisen as an energetically efficient mode of terrestrial locomotion for a small-bodied hominoid moving between arboreal feeding sites'.[16]

This proposition has had its critics, but continues to be quoted with approval. The only thing wrong with it is that it has fallen into what Lewin called the 'teleological trap'. The chimp/human comparison is not comparing like with like. It is comparing the quadrupedalism of a largely arboreal ape with the bipedalism of a species which has been perfecting bipedalism over the space of five or six million years and has transformed its entire skeletal and physiological structure to make it work.

Before bipedalism, the Miocene hominid ancestor would not have had our advantages. Its structure would have differed only minimally, if at all, from that of the other apes with which it was contemporary. It would have been as likely as a modern chimp to find ground locomotion of any kind a comparatively laborious business. For a chimpanzee even *standing up* is a comparatively laborious business. Leslie Aiello comments: 'There is much more muscle activity in a chimpanzee standing upright on two feet than there is in a bipedal human'.[17]

So the ancestral ape would not have walked upright in the anticipation that its far-distant and radically restructured descendants would find it easy. It might have done so if bipedalism was less costly *for itself*—that is, for a single individual anthropoid primate—than quadrupedalism. That is not the case, as Taylor and Rowntree went to a good deal of trouble to find out. It is time we assimilated their message and accepted their advice, that in debates on the origin of bipedalism, arguments hinging on energy costs 'should not be used'.

Fortunately, in practice, most anthropologists have proceeded on the assumption that there must have been some incidental advantage to bipedalism which had nothing to do with the simple process of getting from A to B. They have ceaselessly sought for some other quite different reason for it. And that quest has proved even more fascinating, and even more thickly strewn with casualties.

5

A Surfeit of Solutions

We have, if anything, a surfeit of convincing explanations, rather than a paucity that the AAT uniquely addresses.

C. M. Anderson[1]

Dozens of hypotheses about bipedalism have been found convincing enough to be accepted and published by the peer-review journals over the years. In one sense it is convenient to have so many. It means that anyone who likes a quiet life does not feel obliged, when challenged by the question, 'Why bipedalism?', to pledge allegiance to any one of them. It is easier to wave a hand at the multifarious options laid out like wares in a supermarket, and invite the questioner to 'pick and mix'.

From time to time, with advances in knowledge, it becomes clear that some of them are past their sell-by date, and new ones are brought in to fill the gaps. All those that were based on the belief that bipedalism was a consequence of the hominid's fast-growing intelligence were abandoned as soon as the South African australopithecine fossils were authenticated, combining as they did clear indications of bipedalism with chimpanzee-sized brains. All the hypotheses claiming that bipedalism evolved to facilitate the making and carrying of tools and weapons lost much of their credibility when it was found that the hominids walked erect long before they left any evidence of toolmaking, and long before they hunted big game on the savannah.

The retreat from the savannah paradigm has not resulted in wholesale clearances because it has been a gradual process presented in quantitative terms—a bit more time spent in the shade, a few remnants of climbing behaviour alternating with treks across the grass ... All the same, some of the old favourites are looking distinctly less plausible than they once did.

Scanning the horizon

Sentinel behaviour was a theory that once commanded a big following. It could not fail to appeal to anyone who had seen on television a band of alerted meerkats standing shoulder to shoulder gazing towards the sight or sound of possible danger. They stand as erect as guardsmen, one straight line from tip to toe. Surely our ancestors in their savannah days scanned the horizon in the same way? But bipedalism predated our savannah days.

In any case, even at its best that could only have accounted for bipedal posture, not locomotion. The meerkat, like every mammal given to sentinel behaviour, drops down onto four legs to run away.[2] Not one of the mammals devoted to scanning behaviour has ever taken even half a dozen steps forward on two legs.

Carrying behaviour

Another old favourite was carrying behaviour. At one period it was felt to be almost self-evident that bipedalism evolved to 'free the hands' from the work of participating in the tasks of locomotion so that they would be available for carrying things. Much of the debate centred around what it might have been that they were carrying, since the fossil record indicated that the first bipedalists left no trace of tools or weapons.

C. O. Lovejoy believed they were carrying food.[3] In 1981 he published his theory that the hominids became monogamous, mainly because of the long dependency period of human infants which necessitated some contribution from the male towards the tasks of child-rearing. He suggested that they became bipedal to carry food home to their mates. Lovejoy was one of the small minority of anthropologists who always believed that bipedality predated the savannah, even when that view was pretty heretical (two others were John Napier[4] and Russell Newman[5])—and he has proved to be right about that.

He was not so lucky, however, with the monogamy theory. We have since learnt that when bipedalism first emerged there was no long dependency period. In 1985 research by T. Bromage indicated that the hominid's babies developed quite as fast as the chimpanzee's babies do today.[6] Arguably, too, humanity's degree of commitment to monogamy does not suggest that it has been biologically programmed into us for the last five million years.

There was one more problem about the Lovejoy hypothesis: what kind of food were they carrying? If it consisted of small items like seeds or berries, and if the male were foraging far afield as Lovejoy implied, then their food would have had to be carried home in handfuls and would hardly repay the energy expended on the journey. On the other hand, if it was, for example, meat of some kind, the normal chimpanzee mode of transportation is to walk on three legs and drag it along the ground.

The food-carrying theory was not lightly abandoned, because field workers often report that on the rare occasions when they observe chimps or bonobos moving bipedally—about 1 per cent of the time—it is because they are carrying food. Usually it is the kind of bonanza food set out to lure them to places where they can be easily observed—rare and luscious enough to be tempting, not too big to be portable, and with every incentive to grab a greedy armful and hurry off with it to a safe place. As Harold R. Bauer commented: 'Most of these observations have been made unsystematically in areas which biases the frequency of behaviour (Wrangham, 1974) and overemphasises the importance of bipedal locomotion since wild chimpanzees very rarely carry natural food in a bipedal posture'.[7]

Another popular carrying suggestion is that they were carrying their young. R. E. Sinclair and Mary Leakey suggested in 1986 that bipedalism evolved as a strategy for following migrating herds on the savannah, and that carrying the young bipedally would facilitate this.[8] The necessity for carrying the young in the arms is more likely to have been a consequence of bipedalism rather than a cause of it. It would not have been an improved method of transporting them: in quadrupedal savannah primates such as the baboon, the offspring rides astride the mother's back, leaving her hands free for foraging items of food from the grass. The savannah chimpanzee uses the same convenient method. The Sinclair/Leakey theory merely highlights a basic flaw in the 'freeing-the-hands' model—namely, that it would at best only have freed the hands of 50 per cent of the species. The hands of the female would promptly have been re-enslaved by the necessity of using them to carry the young. In any case, we now know that bipedalism predated the heyday of the savannah and its migrating herds.

A third version of the carrying hypothesis was suggested in 1983, when Leo Laporte and Adrienne Zihlman suggested that in a mosaic habitat the hominids would have had to trek across arid grasslands during the dry season from one patch of woodland to the next, and may have learnt to walk upright to carry food and water with them on these excursions.[9]

'The early hominids may have used containers for carrying water (perhaps initially inside a melon or a root) and for collecting food items, especially the smaller berries and seeds, and the occasional lizard and larval grub.' That description blurs the fact that the first ancestor to attempt bipedalism must have been an ape, not a hominid. An ape's instinctive reaction to berries and seeds and grubs and drinking water is to put them into the container expressly designed for the purpose—its stomach. The l.c.a. would be unlikely to prepare provisions for a distant destination, rationing consumption and considerably slowing its progress by trudging on two legs instead of galloping along on four. As Richard Wrangham and Dale Peterson have shown,[10] the only apes known to dig up roots and carry pieces around to chew on are the chimpanzees of Tongo in eastern Zaire. There is no sign of this behaviour leading to bipedal locomotion.

Keeping cool

Peter Wheeler in the 1980s embarked on an original and prolific series of papers which had nothing to do with sentinel or carrying behaviour. His central theme was that the ape stood up in order to keep cool on the savannah.[11] And he regarded that as a possible explanation of bipedalism. 'The major thermoregulatory advantage conferred by bipedality, to an animal extremely sensitive to hyperthermia, could also account for the initial evolution of this unusual form of locomotion'.

The idea was not entirely new. It had been given a passing mention in 1970 by Russell W. Newman, who wrote: 'Being erect *per se* in open grassland, substantially reduces the solar heat load by minimising the amount of surface area exposed to direct sunlight'.[12] The germ of the idea was even older than that; Newman himself had read it in a paper about sheep in Australia, published twenty years earlier.

However, Wheeler thoroughly researched the cooling effect and made it his own, quantifying the thermoregulatory effect at different times of the day and illustrating it with graphs and photographed models. Newman had treated the thermoregulatory effect as an incidental bonus: 'However,' he wrote, 'the assumption of an upright position was so fundamental to human evolution that this thermal advantage can only be considered as a minor and fortuitous by-product.' Wheeler did not consider it minor or fortuitous, and he supplemented it by comparing wind speeds at ground level with wind speeds at the height of Lucy's head, calculating that the cooling breeze would further contribute to convective heat loss and cutaneous evaporative cooling.

One weak point in Wheeler's hypothesis was the unlikelihood that an animal as excessively sensitive to hyperthermia as he claimed would ever have opted to move out into such an inhospitable environment and to continue foraging when the sun was overhead. At first his calculations were explicitly based on a premise of 'open savannah conditions' or 'open equatorial environments'. By the end of the '80s he had accepted the image of a less stressful microclimate and conceded that the hominid must have rested in the shade during the hottest part of the day to conserve water; but he still held that bipeds would gain a significant advantage by not needing to stay in the shade for as long as quadrupeds of equivalent size.[13]

Wheeler's reasoning was ingenious and eloquent, but as it turned out he was betting on the wrong horse. In their original form his arguments were overdependent on Dart's model of the torrid savannah as the backdrop to emergent bipedalism. In the long run the hominids doubtless did live on the savannah, and the later species could have benefited from their erect posture in the way he describes, enjoying a serendipitous spin-off from their unique mode of locomotion. But avoiding hyperthermia was not the motive which originally caused them to walk erect. We are back with Newman's assessment: these advantages were minor and fortuitous by-products, not first causes.

The long-distance runner

One paper sometimes alleged (though not by its author) to offer an explanation of bipedalism was written by David Carrier in 1984.[14] He asked a new question: 'Quadrupeds may run faster over short distances, but who wins in the long run?' He established that in a very long chase a biped (that is, *Homo*) is able to run down an antelope. One reason is that when a quadruped's front legs hit the ground, the impact travels up the legs and compresses the thorax, and tends to force the air out of the lungs. So it has to synchronise its running with its breathing—one breath per locomotor cycle.

A human biped, on the other hand, is freer to vary its pace. 'They would have been free to pursue prey at any speed within their aerobic range. They could therefore have picked the speed least economical for a particular prey type. This would have forced the prey to run inefficiently, expediting its eventual fatigue.' However, Carrier himself explicitly recognised that the scenario of the long-distance runner could not be used to explain the emergence of bipedalism in the ancestors of Lucy. He suggested it might 'better apply to the evolution of locomotion in *Homo*, and might serve to distinguish *Homo* from the australopithecines'.

Peace-keeping

In 1993 another hypothesis, this time tailored to fit the mosaic model, was suggested by N. G. Jablonski and G. Chaplin. Their paper concentrated on problems that might arise at the woodland/grassland interface as a result of the comparative scarcity of resources. The suggestion was that such a scarcity might result in damaging amounts of internecine aggression, and one way of minimising that would be to increase the use of dominance displays, to deter would-be challengers from harmful and futile attacks.[15]

Displays of breast-beating threat behaviour in gorillas are often accompanied by short bursts of bipedal running; if indulged in more frequently, that might have led to bipedal locomotion becoming commoner. However, this mode of peace-keeping only works if for every threatener there are a number of appeasers keeping their heads down. In a troop in which all the males—to say nothing of all the females—were on their hind legs displaying dominance, such behaviour would not really be conducive to stable social relations.

Feeding behaviour

Some of the bipedalism theories have themselves evolved and interbred to produce versions that are better fitted to survive in the new mosaic paradigm. One such speculation concerns feeding behaviour.

In the 1970s scientists, trying to explain the origins of human behaviour patterns, naturally sought for clues by observing the savannah-dwelling primates, assuming that our own ancestors had been subject to the same pressures. Thus the fiercely macho and hierarchical societies of the baboon *Papio* were treated as possible models of the social life of early Man the Hunter. And for some years a possible hypothesis about the origins of bipedalism was based on the feeding habits of the gelada baboon *Theropithecus*.

In 1970 C. J. Jolly advanced the theory that the first hominids were neither carnivores nor scavengers, but seed-eaters like the gelada.[16] This was considered a possible explanation of human bipedalism. Studies of the gelada revealed that bipedal travel accounted for from 13.6 per cent to 29.6 per cent of the distance covered in a day— a far higher figure than in any other primate except man.

Although it is undoubtedly locomotion and certainly not quadrupedal, the gelada's bipedalism is not a very good model for that of humans. They feed by picking up grass seeds, and to avoid having to get up and walk on four legs from one patch of grass to the next, they shuffle along in a squatting position, supported by their hind legs—and usually also their bottoms—so that the endless hand-to-mouth feeding process can continue without interruption. The process is very slow—about one-third of their quad-rupedal walking speed. The steps taken are necessarily very short because the knees remain bent. And the shuffling is never used for distances greater than a metre—normally, for less than half a metre. Nevertheless, Jolly believed that a similar squatting shuffle in our distant ancestors could have heralded the move to our own erect walking.

R. W. Wrangham in 1980 suggested an improved version.[17] Suppose the harvest was not seeds, but small berries growing on bushes. Because each item was small there would be the same inducement to keep the hands busy picking instead of using them to walk on. But the posture would be nearer to our own. The animals would be standing up, perhaps reaching up over their heads. If the bushes were very close together they might move from one to another *on two legs* instead of going down on all fours. This fits well into the mosaic model, since the forest/grassland ecotone is just where such bushes are likely to be found.

K. D. Hunt tested this conception in a significant field study published in 1994, involving over 700 hours of observation of wild chimpanzees in that type of habitat.[18] It strongly confirmed that chimpanzee bipedalism is predominantly a feeding adaptation. Of the observed instances of bipedal behaviour, only 2 per cent was occasioned by scanning behaviour, 1 per cent by carrying behaviour ('hold infant'), 1 per cent by dominance display, and none at all by locomotion in a non-feeding context. Over 80 per cent consisted of standing up on the ground or in the branches while feeding.

Could this type of adaptation explain why hominids walked on two legs? One difficulty is that the incidence of any kind of bipedalism was small—97 instances in 14,700 chimpanzee hours. But the main snag, as with sentinel behaviour, is that it was in the main merely *postural* bipedalism: only 4 per cent of terrestrial bouts involved locomotion, and these were limited to short-distance within-site shuffling. Also, the bipedal posture was usually assisted by a forelimb holding on to a branch to aid stability, thus involving three limbs or, at the lowest estimate, two and a half. Hunt concluded, 'A bipedal postural feeding adaptation may have been a pre-adaptation for the fully realised locomotor bipedalism apparent in *Homo erectus*'.

As a pre-adaptation it works perfectly. After its publication Bernard Wood, in *Nature*, went a step further and suggested that the truth may lie in a merger of Hunt's hypothesis (accounting for developments leading up to the verge of bipedalism) and Wheeler's thermoregulatory one (explaining advantages that followed its emergence).[19]

But that still leaves a yawning gap: What came in between? What carried one section of the ancestral apes over the verge and left the chimpanzees and the gorillas stuck at the 'pre-adapted' stage for the next five million years? It does suggest the possibility that something must have happened to our own ancestors which did not happen to theirs.

It has been argued that with all this wealth of possible explanations to choose from, the last thing we need is yet another hypothesis. On the other hand, there is no sign as yet of an emerging consensus. It would be hard to find a scientist willing to select one of the above and announce, 'This is obviously the solution. It explains everything. The case is now closed.'

6

The Wading Ape?

It seems to me likely that Man learned to stand erect first in the water and then, as his balance improved, he found he became better equipped for standing up on the shore when he came out.

Sir Alister Hardy[1]

In the past, bipedalism theories were based on the supposition that the first bipeds lived on the savannah. Currently, they are based on the supposition that the first hominids lived in an environment halfway between woodland and grassland.

It might be instructive to take a specific case, the case of the Ethiopian fossils—that is, Lucy, and the group of fossils of mixed sexes and ages that came to be known as the First Family. What kind of place was it when those hominids lived there?

Hadar

The palaeoenvironment was described in considerable detail in a paper published in 1982 by Donald Johanson, Maurice Taieb and Yves Coppens.[2] 'Generally,' they reported, 'the sediments represent lacustrine, lake margin and associated fluvial deposits related to an extensive lake that periodically filled the entire sedimentary basin.'

Descriptions of the various strata describe the palaeoecology at successive geological periods, in terms of marsh and flood-plain deposits, shallow water lacustrine conditions— a second lacustrine incursion—mini-delta formations—a lake surrounded by marshy areas—fairly rapid fluctuation in lake level—expansion of the basin with fully lacustrine conditions. At site AL333, where the First Family was found, surprise was expressed that virtually no specimens of other species were present 'except for fish and rodents'. To the savannah, as we used to think of it, there is only one reference: 'Undoubtedly episodic regressions opened up grassy areas, but always with some bushy to wooded areas along streams.' ('Undoubtedly' here has something of the ring of Klein's 'surely'.)

The overall impression given is that Hadar when the hominids lived there was a pretty soggy place.

In 1983 the first scientific meeting held by the Institute of Human Origins in Berkeley featured a quite heated debate on the Lucy fossil and what her skeleton told us about her mode of life. One school of thought held that Lucy was a fairly committed terrestrial bipedalist even though, with her shorter legs and longer feet, her gait would have been more laborious than ours. An opposition faction argued that although many features clearly indicated bipedal adaptation, evidence from her fingers and toes showed that she still spent a good deal of her time in the trees.

There was no disagreement about the nature of the habitat. Both sides shared the assumption that there was terra firma for Lucy to walk on, and there were trees—trees big enough for an animal of her size to climb in if she was so minded. The only difference was in the relative amount of importance they attached to those two respective features of the habitat.

There was also the lake, to which neither side attached any importance. Periodically, as we have been informed, the lake spread out until it engulfed the whole basin, its waters lapping around the trunks of the trees and submerging the terra firma.

There must have been times in such areas when Lucy's ancestors were unwilling to forgo the food supplies still visible on the branches of the partly submerged trees. Seeking to exploit them would have placed them in the one situation where an ape, still largely arboreal, would be obligatorily bipedal as soon as it descended to ground level.

It has often been observed that bipedalism involves considerable costs to set against the many advantages it nowadays confers on *Homo sapiens sapiens*. In the earliest unhabituated stages the costs would have been at their highest and most of the advantages still in the future. Arguably, for this reason, it would never have been resorted to except under duress, and that would explain why our way of walking is unique. Flooded terrain could have constituted the duress.

If we postulate that bipedalism arose as a consequence of wading behaviour, we are in no danger of falling into the teleological trap. For an anthropoid ape in three feet of water, the motive for walking upright—however clumsily and laboriously—does not lie in some advantage that might accrue to its descendants. It is immediate and individual and, indeed, indispensable. The advantage is that it allows the animal to go on breathing, whereas if it walked on four legs its nostrils would be under water.

The proboscis monkey

There is one monkey which lives exclusively in a habitat consisting of a mixture of trees, land, and water, and that is the proboscis monkey. It lives in mangrove swamps on the coasts of Borneo in an area interlaced with streams, and when the tide comes in the proboscis monkey often finds it impossible to move from one feeding ground to the next without crossing water.

It regularly crosses quite wide stretches of water on two legs. For females with young this mode of progress is obligatory, even when the water rises high enough to make it possible to swim across.[3] Unlike the young of terrestrial primates, the infant clings to its mother's chest rather than riding on her back. If its mother attempted to swim while carrying it, it would drown.

Proboscis Monkeys. *Amanda Williams*

This has not led to the proboscis monkeys as a species becoming habitually bipedal, but they do resort to occasional bipedalism on land as well as in water. What distinguishes this from the occasional bipedalism of other primates is that it is not hylobate-type bipedalism—they are perfectly capable of efficient quadrupedalism—and it is not connected with sentinel behaviour or dominance behaviour or feeding behaviour. It is merely an alternative locomotor mode of getting from A to B. It is also unusual in that it is sometimes group behaviour, with the monkeys following one another bipedally in single file.[4]

Chimpanzees

It is true that a proboscis monkey is a monkey and not an ape, and therefore a fairly distant relative of ours. Behaviour of this kind has not been observed in common chimpanzees in the wild, but the obvious reason for this is that the chimpanzee does not live in mangrove swamps.

There is, however, an interesting little table drawn up by H. R. Bauer in 1977.[5] In it he listed the different settings in which bipedal behaviour has been observed in the chimpanzee. It is interesting to note that almost all these occasions of bipedalism correspond to theories about the origins of bipedalism—sentinel behaviour, dominance behaviour, carrying behaviour, feeding behaviour. Only number 5 has been overlooked—the fact that the chimpanzee walks on two legs if it can thereby avoid getting its front legs wet.

The third African ape, the bonobo, was only discovered in this century. Formerly known as the pygmy chimpanzee, it has now been recognised as a separate species—*Pan paniscus*, as distinct from the common chimpanzee, *Pan troglodytes*. The first detailed description of it was given in 1933 by Harvard zoologist Harold Coolidge. It is regularly described as 'the most human' of the apes—a particularly lively, intelligent, graceful and peace-loving animal.

Table 3. Cases of observed bipedal standing or movement (from H. R. Bauer, Chimpanzee bipedal locomotion in the Gombe National Park, *Primates* (1977). **18**(4), 919).

Setting	Physical situation
1. Standing, looking towards other members of a travelling group of chimpanzees, which may or may not be foraging (VAN LAWICK-GOODALL, 1968).	Rocks, tall grass, or other vegetation blocking ground view.
2. Standing, looking towards buffalo snakes, strange chimpanzees on boundary patrols, and, once, in response to a leopard-like sound.	Rocks, tall grass, or other vegetation blocking ground view.
3. In agonistic displays directed at chimpanzees or baboons (VAN LAWICK-GOODALL, 1968), standing or moving.	In grass, on the forest floor, or on limbs of trees.
4. Picking up provisioned food and carrying it in arms (VAN LAWICK-GOODALL, 1968).	In and around a provisioning area.
5. Walking over wet ground (VAN LAWICK-GOODALL, 1968).	During or after a rain.
6. Walking or standing on tree limb briefly, sometimes foraging.	Normally when there are no nearby branches.
7. With injured forelimb, standing, walking, or running is more common, bipedally.	Symptoms of injury may be swelling and avoiding contact with objects.

A paper was published in 1978 entitled 'Pygmy chimpanzees as a possible prototype for the common ancestor of humans, chimpanzees and gorillas', written by Adrienne Zihlman, J. E. Cronin, D. L. Cramer and V. M. Sarich. It drew attention to bonobo/human resemblances such as the proportion of the limbs, the narrower trunk, relatively small canine teeth, and the fact that in captivity bonobos walk bipedally more often than common chimpanzees do.[6]

Randall L. Susman, who has consistently argued that Lucy was largely arboreal in her lifestyle, visited the bonobo in the wild and supported the idea that studying it might be more relevant to our own origins than studying the savannah chimpanzee. Susman's paper in 1987 stressed the point that '*Pan paniscus* is the most forest adapted of the African apes'.[7]

One thing was rather surprising about this flurry of interest in the bonobo. No one specifically posed the question of exactly *why* humans resemble the bonobo more closely than they resemble the other African apes. In popular summaries the position was apt to be summed up as: 'The bonobo is regarded as our closest living relative'.[8]

Whether it is *genetically* our nearest relative is dubious. Most systematists believe that the hominid and the chimpanzee lines split apart before the chimpanzee line divided into two. If any bonobo/hominid resemblances are due to anything other than sheer chance, it must be because of selective pressures in the environments in which they lived. In simple terms, something happened to the ancestors of bonobos and humans which did not happen to the ancestors of chimpanzees and gorillas.

It could not have been simply a matter of living in forested areas. All the gorillas and most of the populations of chimpanzees also lived in forested regions. There must have been something different about the bonobo's forest.

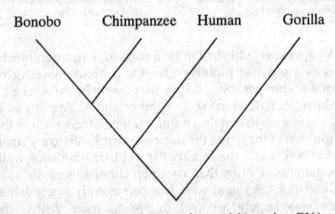

Measurements of relatedness based on haemoglobin and on DNA-DNA hybridisation both indicate that the bonobo is no more closely related to us than is the common chimpanzee. (Morris Goodman and C. G. Sibley, in *Camb. Enc. Hum. Evol.*)

Frans de Waal in 1989 appears to have been the first who specifically focused attention on the difference:[9] The forest areas which form the core of the bonobo's habitat in Zaire are seasonally flooded. In the past, when Africa was wetter than it is now, the flooding may well have been greater in extent, or in annual duration, or both. Everyone visiting the site describes the difficulties of approaching it even at a favourable time of year—'. . . a day's travel by motorised dugout canoe . . . and finally on hour or so's hike through the swamp',[10] or 'an eight-day riverboat ride; a two-day trip by Land-Rover; and a day's hike through swamp'.[11] None of the scientific papers had made a point of the fact that the bonobo lived in swamp forest areas.

Frans de Waal treated the water factor as interesting and possibly significant: 'Bonobos are remarkable in that they do not fear water . . .' [They] 'are known voluntarily to enter pools or moats, splash in the water and even dive completely under'. He quoted accounts of female bonobos '. . . walking upstream, in the water. They snatched handfuls of floating dead leaves, picking out things to eat'.

Frances White observed the same thing: 'One thing we share with bonobos but not common chimpanzees is a love of water. Especially on hot days in the dry season, a party of bonobos will spend most of the day walking up and down the shallow, fast moving streams searching for edible roots, insects, and even small fish, which they catch by scooping debris on to the sandy banks'.

Also: 'Susman,' de Waal wrote, 'has observed that the numerous bonobo tracks along stream-beds lack knuckle prints. This suggests to him that bonobos avoid getting their hands wet by assuming bipedal postures when crossing streams. So Hardy's aquatic ape theory, or at least the part that links bipedalism to wading in shallow water, may help explain why bonobos have such strong, long legs'. A photograph of two bonobos taken by de Waal illustrates the long, strong legs, and the stance more erect than that of the common chimp.

Gorilla

There was one photograph in the *National Geographic* in 1995 of a gorilla using bipedal locomotion not for display behaviour, or sentinel behaviour, nor for reaching up for food on an overhead branch. It was walking bipedally in order to get from A to B. It was up to the top of its thighs in water, and it seems clear from the photograph that it had already 'sloshed along', as photographer Michael Nichols described it, for quite a considerable distance.[11] The caption accompanying the photograph read: 'While it's not news that sedges serve as a starchy supplement to a gorilla's diet, biologists have only recently discovered that the apes will go wading to get them'.

Adding to the surfeit

The idea that bipedalism may have arisen as a consequence of wading behaviour is a hypothesis, just like all the others outlined in the previous chapter. None of them is proven. Indeed, it could be argued that the multiplicity of suggested solutions is itself a clear sign that none of them has been found satisfactory. There is this to be said in favour of the aquatic hypothesis: it is not merely a one-off answer to the one-off question of why we walk on two legs. It provides a scenario which at the same time offers a possible solution to a wide variety of other unsolved problems.

The next of these is: Why naked?

7

The Naked Ape

No one supposes that the nakedness of the skin is of any direct advantage to man.

Charles Darwin[1]

In most of the attempts to account for the emergence of the naked ape, the nakedness has been found even harder to explain than the bipedality. When a problem has remained unsolved for as long as these two, the possible responses are either to intensify the efforts to find a solution, or to abandon them as unprofitable and concentrate on something else. Generally speaking, scientists have made the first response in respect of bipedalism and the second in response to the loss of body hair.

There is certainly no surfeit of solutions to the problem of why the naked ape became naked. Seven years ago, in *The Scars of Evolution*, I gave examples of otherwise comprehensive textbooks on physical anthropology in which this feature was dealt with in literally three words, or not at all. There are no signs that this attitude is about to change. In 1992 *The Cambridge Encyclopedia of Human Evolution* gave an excellent and updated account of the new thinking that was going on in all the related fields.[2] The fact that it did not mention hairlessness simply reflects the fact that in that field nothing is going on.

From time to time, however, the problem still raises its head, despite a notorious attempt earlier in the century to exorcise it for good, by arguing that we are in fact hairier than chimpanzees. More specifically, human hair follicles were said to be closer together than chimpanzee hair follicles. 'The assertion that we are the least hairy of all the primates is, therefore, very far from being true; and the numerous quaint theories that have been put forward to account for the imagined loss of hair are, mercifully, not needed.' (The last time I quoted that I did not know its origin; I regret to report that it was first written by the otherwise admirable Frederick Wood Jones in *Man's Place Among the Mammals*.[3])

Some earlier examples of the 'numerous quaint theories' were recapitulated in *The Scars of Evolution*. One was a suggestion that hairlessness helped to get rid of insect parasites. Anther proposed that the males in one sub-population of the l.c.a. acquired a kinky hankering for bald-bodied mates, so that all the females lost their body hair by a process of sexual selection. However, the aim here is to consider the ideas that have remained in contention since around 1970, and to identify the different strands in their reasoning. One of the issues that divides them is (as with bipedalism) whether the hominids lost their hair in the forest, or after they moved out of it.

Naked in the forest?

Russell Newman suggested the first possibility in his challengingly entitled paper, 'Why man is such a sweaty and thirsty naked mammal'.[4] From the beginning he treated nakedness as a sad loss for the species:

> One of the most important mammalian defences against radiant heat loss is a dense and highly reflective coat of body hair. This serves multiple purposes: reflecting up to half of the solar energy, absorbing some of the unreflected portion at a distance from the skin for dissipation by convection and re-radiation, and providing an insulative space against the conductance of the absorbed heat towards the skin.

He points out that a dense coat is conspicuous in large desert animals otherwise in danger of overheating by day and hypothermia by night. 'Man's present glabrous state,' he asserts, 'is a marked disadvantage rather than the other way round' and therefore 'loss of hair must have stemmed from other causes or preceded the occupation of the habitat in question [that is, the savannah] at least for its inception.' The time and place when it must obviously have happened, he concludes, is the ancient forest habitat.

His case is competently argued and convincing. The only let-down is that he does not answer his title question: 'Why?' If hairlessness stemmed from 'other causes', what were they? He reaches the conclusion that the forest is the place where progressive denudation would have been *least disadvantageous* for the ancestors of the hominids. In that case it would also have been least disadvantageous for the ancestors of the African apes. Absence of disadvantage is not a sufficient reason to discard a feature which has been a shared inheritance of all primates and most mammals since they first appeared on earth.

Bigger means balder?

In 1981 Gary Schwartz and L. A. Rosenblum—who also rejected the idea that nakedness evolved in savannah conditions—indicated a possible reason why it might have been initiated in the forest. They reported a negative correlation between hair density and body size in primates.[5] Larger ones, like the great apes, have fewer follicles per square centimetre than smaller ones like monkeys. They speculated that 'the allometric trends may betray clues to the evolution of human hairlessness' and that merely as a result of growing larger 'a substantial depilation occurred on the basis of body surface area some time prior to man's migration from a forest to a grassland habitat during the Pliocene'.

It was not entirely an original idea. Adolph A. Schultz

had commented on relative hair density in primates in 1931,[6] and concluded that man's hair loss is 'merely the most extreme manifestation of a general evolutionary trend to reduce the hairy coat of the largest primates'.

There are two snags. As with Newman's account, this one does not predict any different sequels as between the apes and the hominids. An adult male mountain gorilla is larger than a man and, according to this theory, should have sparser hair than *Homo*, but has in fact a thick and lustrous pelage. The chimpanzee's coat appears sparser, but the sparseness can be overestimated by studying only bored and captive specimens. Both chimpanzees and bonobos in these circumstances are liable to develop the habit of plucking the hairs from their arms, rather as some caged parrots pluck their feathers out.[7]

The second snag is that human functional nakedness is not caused by having the hair follicles more widely spaced. It is caused by having follicles that produce hairs so short and fine that they are not visible to the naked eye: some are too short to reach the surface of the skin. The evolution of this condition is likely to have consisted of a progressive shortening of the body hairs, sustained over a long period. There seems to be no trend in the primates towards shortening of body hair, whether in response to body size or any other variant.

Cool on the savannah?

Peter Wheeler took the view that the nakedness combined with bipedalism to provide a defence against overheating on the savannah.[8]

This sounds intuitively right: who would wear a fur coat at midday in the tropics if it was possible to take it off?

But it is not that simple. Russell Newman was equally convinced that a naked skin in the tropics is a sure way of getting even hotter. Experiments had been conducted by shaving the hair from various animals and seeing how they fared. Newman reported: 'There is no evidence that a hair coat interferes with the evaporation of sweat; what can be said is that exposure to the sun after removal of the body hair increases sweating in cattle (Berman, 1957) and panting in sheep (McFarlane, 1968) because the total heat load has increased'.[9] Other studies he quoted showed that at rest or during light work, a clothed man gains about two-thirds as much heat as a nude man and sweats commensurately less.

Newman's thinking also seems intuitively right. What Bedouin would set out across the desert without a burnous to shade him from the sun?

Wheeler's case, like Newman's, was ingenious and well argued. Unfortunately for him, he had welded the bipedalism and the nakedness tightly together as an inter-dependent response to savannah conditions. When it became clear that bipedalism long preceded life on the sav-annah, the whole structure suffered some loss of credibility.

Primate skin

The scientist who made the most determined and sustained attempt to solve the problem was William Montagna.

He conducted intensive, exhaustive and detailed researches into all the aspects of primate skin in a large number of species, looking for clues, and published numer-ous papers on the skin of primates in general, the skin of chimpanzees and gorillas in particular, and the evolution of human skin. These publications added greatly to our understanding of the whole subject. In 1972 he wrote:

Although this approach expanded my own knowledge of cutaneous structure and function, it failed to explain the unique feature of man's skin—its almost complete nakedness. Since it is this single factor that constitutes the chief difference between human skin and the skin of other mammals, we are left with the major objective of our study still unattained.[10]

Having failed to answer it, he did not succumb to the temptation of belittling it, or suggesting that the difference between ape and human is only quantitative, and after all not very great. It remains a scientific Everest, it is still unscaled, and there has to be some reason why Montagna's objective was unattained. It is perfectly sensible for scientists to bypass this area of inquiry if they see little hope of making progress in it and wish to concentrate on more accessible problems. What is unacceptable is to respond to the lack of progress by dismissing it as a minor issue, or conjuring it out of existence by sleight of hand as Wood Jones tried to do.

8

The Other Naked Mammals

Another most conspicuous difference between man and the lower mammals is the nakedness of his skin. Whales and porpoises (Cetacea), dugongs (Sirenia) and the hippopotamus are naked.

Charles Darwin[1]

Motagna's experience suggests that the explanation of human nakedness cannot be found by comparing humans with other primates. A different approach is needed. A possible one is to ask the question which Desmond Morris propounded in *The Naked Ape*: 'Where else is nakedness at a premium?'[2] It was a very good question, even though in that book he never got around to answering it. Although the species *Homo* is the only habitually bipedal mammal in the animal kingdom, it is not the only naked one. It is strange that the papers in the professional journals do not seem to regard that fact as potentially relevant to our condition.

At the present time the trend in scientific education does not encourage that line of enquiry. The growth area is in the important fields of systematics and taxonomy—the attempt to establish precisely the evolutionary relationships between different species. In order to construct their family trees students are trained to concentrate on and evaluate all characteristics that are shared between related species—plesiomorphic characters. For them, these are the vital clues. Any feature of any species which is not shared by any of its nearest kin is classed as 'autapomorphic'—a one-off, an idiosyncrasy—and therefore useless, arbitrary and irrelevant to the task in hand. To the more committed

77

systematicists, betraying a keen curiosity about an autapomorphic trait is a sign of scientific illiteracy: it means the questioner has failed to grasp the principle that these one-off aberrations are meaningless and *do not tell us anything*.

It is true they tell us nothing about systematics. Our nakedness, for example, tells us nothing about the primate tree or how we fit into it. That is why we can never explain it by studying only the primate family tree. What an autapomorphy can sometimes tell us is what happened to a particular species during and/or after the time when it was separating from the nearest related species. Sometimes it can tell us why the separation took place.

Consider the example of one particular trait. Construct six cladograms setting out the family trees of six widely separated classes: (1) bears; (2) mustelids (stoats, weasels, etc.); (3) game birds; (4) canids (dogs and their relations); (5) lagomorphs (hares); and (6) cricetidae (lemmings).

In the six cladograms, at the end point which lists all the extant member species, it will be noted that at least one species in each diagram is either seasonally or permanently white. They are the polar bear, the ermine, the ptarmigan, the Arctic hare, and the Arctic collared lemming. The canids have two—the Arctic fox and the Arctic wolf.

In every case the whiteness is a non-derived characteristic—an autapomorphy. But it is not meaningless that it crops up so many times. It enables us to guess what happened to these animals during or after speciation and what caused the speciation. Possibly the species moved north into the Arctic circle. Possibly in one of the ice ages, arctic conditions spread south and forced it to adapt, and subsequently it retreated northwards following the conditions to which it had become specialised.

It takes little or no imagination to speculate why the colour change took place—because white is the colour of snow. It is a kind of convergence. The term convergence is most often used in terms of two unrelated species coming to resemble one another in respect of a whole cluster of features, including general body structure. Examples are dolphins adopting the same outline as fish, or insectivores from the different continents separately evolving similar silhouettes, with powerful front limbs and long narrow tapering snouts. But the Arctic fox and the Arctic hare and the others have converged in respect of one single trait only.

This conclusion is generally accepted. Nobody argues that it is invalid on the grounds that not all creatures with white pelage are necessarily Arctic or ex-Arctic. There are white egrets and white cockatoos, white horses and white cattle. And it is not rejected on the ground that some residents of polar regions—like the penguins—have not turned white.

It is not difficult to see the possibility of an analogy between the evolution of whiteness in an Arctic environment, and the loss of body hair in an aquatic environment. Nearly all extant naked mammals are in some degree aquatic. The exceptions are the naked Somalian mole rat which never comes to the surface; the group of animals which used to be known as pachyderms; and humans. It is true that by no means all aquatic mammals are naked: below a certain size (roughly the size of Lucy) none of them are. But there is as much reason to regard nakedness as one adaptation to an aquatic habitat as there is to regard white or seasonally-white fur or feathers as one adaptation to an Arctic one. As with whiteness, it is quite possible to think of reasons why this should be the case. Fur is a far less efficient insulator in water than it is in air. Scientists like Scholander[3] and Sokolov,[4] when writing about the hairless aquatics, have frequently noted the advantages of trading the coat of hair in for a layer of blubber.

The 'pachyderms' are an interesting group. If it were possible to subtract them from the equation, then the naked non-aquatic mammals would be reduced to two species: *Homo sapiens* and the naked Somalian mole rat—very unlikely bedfellows.

Pachyderms. *Amanda Williams*

'Pachyderm' (from the Greek 'thick-skinned') was a word used by zoologists in the last century to refer to a group of animals which seemed to them to fall naturally into a sub-set of mammals; it included such animals as the elephant, the hippopotamus, the rhinoceros and the walrus.

The term is still sometimes used by laymen as a facetious word for elephant, but scientists nowadays very rarely use it. They know there is no taxonomic relationship connecting these animals with one another. The rhinoceros is a perisso-dactyl related to the horse and the tapir; the hippopotamus is an artiodactyl related to camels, cattle, sheep and giraffes; the elephant comes from a very ancient lineage and has no living relatives, though it is thought to be distantly connec-ted with the sloth, the hyrax and the dugong. Students of systematics tend to feel that any apparent affinity between the extant pachyderms which may have been imagined in the past must have been in the eye of the beholder.

It should not be assumed, however, that the beholder's eye was untrained. The word was coined by Georges Cuvier, the man generally credited with establishing com-parative anatomy and palaeontology as sciences. He saw a number of large, thick-skinned animals, hairless or sparsely haired, and detected or imagined some kind of affinity between them. None of his contemporaries seemed to doubt that there was a class of mammals to which this term could properly be applied, although there was not always agreement about exactly which species were actual members of 'the pachydermous tribes'. Most of them recog-nised the walrus and the pig as legitimate contenders. One man even voted for the giraffe, claiming that its skin was over an inch thick.

To a modern taxonomist these discussions seem as archaic as the debate about whether or not the porpoise is a fish (a proposition once hopefully defended by Christians who wanted to eat it on Fridays). The classification 'pachy-dermata' was factitious, manifestly a blunder. And yet, if we try to look at the pachyderms from the naïve standpoint of pre-Darwinian science, it is easy to see why biologists felt the need of a word for them.

It is not only a thick skin that they have in common: there is a cluster of resemblances. They are all ponderous animals, heavier than other mammals of comparable length and height. Their bodies tend to be barrel-shaped. Many are naked. Their skins tend to be lined with a layer of fat.

They tend—even the most terrestrial ones—to be unusually good swimmers. There used to be an old wives' tale to the effect that pigs cannot swim because when they tried they cut their throats with their trotters. A. R. Wallace commented on this in *The Malay Archipelago*:[5]

> There is a popular idea that pigs cannot swim, but Sir Charles Lyell has shown that this is a mistake. In his *Principles of Geology* (10th edition, vol. ii, p. 355) he addresses evidence to show that pigs have swum many miles at sea and are able to swim with great ease and swiftness. I have myself seen a wild pig swimming across the arm of the sea that separates Singapore from the Peninsula of Malucca, and we thus have explained the curious fact that, of all the large mammals of the Indian region, pigs alone extend beyond the Moluccas and as far as New Guinea.

Elephants do even better. It has long been observed that, like the pig, they are capable of crossing between islands in the East Indies. In a documentary film (BBC, 8th August, 1996) a professional 'tusker tamer' called Randall Moore reported an incident of a group of animals completing an eight-hour swim across the lake at Kariba. (Elephants live a long time and are reported to have long memories. It is just possible that they were being led along an infrequently used route which had been far less arduous before the dam was built.)

Also, palaeontologists have recently discovered that

ancestors of the elephant crossed from North to South America ten million years ago, millions of years before a land bridge existed between the two land masses, and millions of years before any other species is known to have crossed the gap.[6]

We are looking, then, at a cluster of autapomorphic characteristics which (like the whiteness of Arctic animals) has evolved in a number of widely separated species, bestowing on them the superficial resemblances which misled Cuvier.

Among the terrestrial pachyderms, the degree of hairlessness varies. Some of them have—or have had—hairier relations. There used to be a woolly mammoth and a woolly rhinoceros. The domestic pig has little or no functional hair, but some of its wild relatives, like boars and warthogs, have retained their pelage. Or perhaps, rather than retaining it, they re-acquired it. The coat they now have consists of unusually bristly hairs, with no heat-retaining underlayer, apparently designed less for thermoregulation than for damage limitation when charging through the undergrowth. On the other hand, one Asian wild pig—the babirusa—is completely hairless.

It is reasonable to suppose that there may be some good explanation of why the pachyderms resemble one another. There may have been some particular kind of environmental pressure to which their ancestors were—at times and places probably very far apart—all separately subjected. The likeliest guess is that it must have been something to do with water.

This assumption is beginning to command general acceptance. The grouping which includes the elephants, sirenians and hyraxes is known collectively as the Tethytheria. In 1993 M. S. Fischer and P. Tassy[7] wrote: 'We assume a semi-aquatic ancestor for the Tethytheria' and quoted C. M. Janis[8] in support of their conclusion that '. . . in the Tethytheria there was a trend to a semi-aquatic or even aquatic existence'. In the same volume as the Fischer/Tassy paper,

V. M. Sarich[9] points out that whales are 'very likely' to be a sister group of the hippopotamus. (And the hippopotamus is related to the pig.)

Some pachyderms—walrus, dugong, hippopotamus, and so on—are still wholly or predominantly aquatic. The tapir lives in a riverine habitat and the babirusa is a marsh dweller. The terrestrial ones like the rhino and the elephant take every available opportunity to wallow in water or plaster their hides with mud, as if their skins were still uncomfortably ill-adapted to constant exposure to sun and air.

Apparently the only reason that has been suggested for their nakedness is their size: big animals have a lower ratio of skin surface to body mass and may have more trouble getting rid of surplus heat. But there cannot be a single causal connection between the two: a bison is nearly eight times as big as a pig but has kept its thick protective coat.

While we are on the question of size, why are most of them so ponderous? All that extra weight is very expensive for a terrestrial mammal and onerous to carry around. The carefree pachyderms are the aquatic ones. The walrus in water is as lithe and graceful as a ballerina. The hippopotamus paces the bed of the river as lightly as a moon-walking astronaut. It saves so much energy by spending its days in water that it only needs to consume *half* as much food, relative to body mass, as a terrestrial grazer. The same question of ponderousness arises with some non-mammals. Can anyone believe that natural selection cursed the Galapagos tortoises with those excruciatingly cumbersome shells on *land*, where there were no predators around to threaten them? It is much easier to think of them as grounded turtles. Water is the only environment on this planet where it costs nothing to be built like a tank.

If we compare human skin with that of the pachyderms, it has a number of unusual features in common with them, including the following:

(a) Our skin over most of the body surface is virtually naked.
(b) It is substantially thicker than that of any other primate.
(c) It has a fatty layer under the skin considerably thicker than that found in any other primate.
(d) 'Unlike the skin of other primates, the outer surface of human skin is criss-crossed almost everywhere by fine intersecting congenital lines. This characteristic feature is not limited to human skin; creases are also found in the skin of pachyderms, walruses and, to a lesser extent, pigs.'[10]

All these are non-derived characteristics, not found in any other primate. They can all be found in either some, or all, of the pachyderms.

That does not mean that the aquatic ape would ever have been classified among the pachyderms—its acclimatisation to water was relatively brief. But it *would* mean that only two types of environment are known to have initiated hairlessness in mammals—an aquatic one, or a 100 per cent subterranean one.

It would make it that much harder to delude ourselves that humanity's ancestors, alone in the animal kingdom, lost their hair by wandering out into the sunshine.

9

The Fat Primate

A thick fatty layer is as characteristic an attribute of human skin as it is of pig skin.

William Montagna[1]

Why are humans so fat?

We are proud of being more upright and smooth skinned and large-brained and dexterous and eloquent than our nearest kin in the animal world, but we are not proud of being fatter. Yet there can be no doubt that it is a species-specific human feature as distinctive as bipedalism. And the fatness cannot be dismissed as a characteristic acquired in our lifetime, due to greed or wrong food or insufficient exercise. It is present in our babies in even more remarkable quantities than in their parents. Their chubby cheeks and rounded limbs are among the markers that help to trigger parental protective processes. A. H. Schultz has described how different they are from infant monkeys and apes which '. . . resemble in their faces emaciated and toothless old men'. He wrote:

> The average weight of the human new-born is usually given as 3.2 kg., or far more than the corresponding figure for any other primate. This discrepancy is partly due to the fact that most human babies are born well padded with a remarkable amount of subcutaneous fat, whereas monkeys and apes have very little, so that they look decidedly 'skinny' and horribly wrinkled.[2]

The baby's fatness in relation to its size begins to manifest itself in the thirtieth week of pregnancy and continues increasing for the first year after birth. Keeping such a baby alive demands a mother with physical resources capable of gestating and feeding it. Rose Frisch stated that if a woman's adipose tissue constitutes less than 17 per cent of her body weight she ceases to menstruate and cannot reproduce.[3] Yet many wild animals remain perfectly healthy and reproductively fit with less than 2 per cent body fat. At some point in the evolution of our own species natural selection clearly began to lay down high minimum requirements in this respect, which do not apply to other primates. The average human contains ten times as many adipocytes (fat cells) as would be expected in an average mammal of similar size.

We have no way of ascertaining at what point in our evolution humans began to get fatter. But contrary to some conjectures, it cannot have been after the introduction of agriculture led to a more settled existence and more carbohydrates in the diet. The famous Venus figures from the Palaeolithic period are sufficient proof of that. Making all due allowance for possible exaggeration and artistic licence, we must conclude that the sight of a fat woman was not outside the experience of whoever made these carvings.

To keep us warm?

'It might appear curious,' Peter Wheeler observed in 1984, 'to lose a covering of body hair to facilitate heat dissemination, only to replace it with a layer of fat to prevent this loss'.[4]

It does appear curious. But that was the commonest hypothesis in terms of the savannah scenario. Wheeler explained it by drawing attention to the fact (often glossed over in discussions about nakedness) that although the savannah is a very hot place during the day, it can be very cold at night. He argued that the fat layer could have evolved to keep the hominids warm at night without being a serious

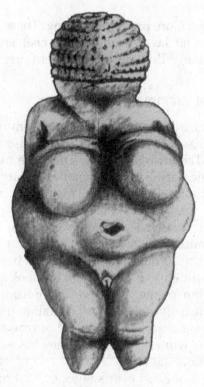

The Venus of Willendorf. Palaeolithic figurine found in Austria. *June Peel*

deterrent to heat loss during the day, because it could be bypassed by capillaries in the skin and did not interfere with the evaporation of sweat.

However, it would have had a number of serious drawbacks which he did not include in his calculations. Both for hunting predators and for fleeing prey, the extra weight would be a handicap in a savannah environment by slowing them down. Producing and maintaining the fat layer would be far more costly in energy terms than maintaining the hair it was allegedly replacing. And the extra energy consumed in carrying the extra flesh around all day would itself increase the danger of overheating the body and necessitate even more profuse sweating. There was no way of making the fat layer sound like a neat answer to the hominid's putative difficulties with thermoregulation.

The other fat mammals

Since this feature is clearly not inherited from the primates, the AAT approach is to enquire: 'Where else is fatness at a premium?' The answer is clear. There are two classes of mammals which are liable to accumulate large quantities of adipose tissue—hibernating mammals and aquatic mammals.

Hibernating mammals include a wide variety of species—dormice and hedgehogs and polar bears—and many of them creep into holes in the earth before going to sleep for the winter. V. E. Sokolov in his book *Mammal Skin* wrote about burrowing animals: 'The dermis of shallow burrowers is usually rather loose, containing great quantities of fat by autumn. Many shallow burrowers hibernate throughout the winter. By autumn they have accumulated a vast, continuous fat layer.'[5]

However, no one has ever suggested that our ancestors crept into burrows and slept away part of the year. Besides, the fat of burrowing animals is different from ours in several respects: it does not adhere to the skin, it has a high content of unsaturated acids and a very low congealing temperature, and it is only present at certain seasons. Some non-burrowing animals which need to store fat to cope with seasonal scarcity tend to accumulate it in specific sites, like the fat-tailed sheep—but these again are obviously not good models for what has happened in the case of *Homo sapiens*.

The closest parallel, then, is with the aquatic mammals. Most of the marine mammals—and indeed also that most marine of birds, the penguin—have a thick layer of fat under the skin. This tendency is not confined to sea mammals. Smaller freshwater aquatics like the beaver typically have higher proportions of body fat than their nearest non-aquatic relatives. It is true that one article in the *New Scientist* stated in respect of the hippopotamus that contrary to appearances '... it is certainly not fat'.[6] But this statement was made by a man who was trying to promote the marketing of hippopotamus meat. Perhaps what he meant was that a hippo steak would be less marbled with fat and contain less cholesterol than a beef steak. But the fact remains that the skin of the hippopotamus is lined with a layer of fat 5 cms thick.[7]

It is generally assumed that the purpose of the fat layer in aquatic mammals is thermoregulation. They would have had a more urgent reason for acquiring it than savannah hominids, because water conducts heat away from the body much faster than air does. Experiments were conducted comparing the relative efficiency of the fat layer as compared with a coat of hair in maintaining body heat in water. In 1950 P. F. Scholander and his colleagues reported that while a coat of hair provides the best insulation in air, a layer of fat gives the most efficient protection against heat loss in water.[8]

V. E. Sokolov shared the assumption that aquatic mammals acquired the fat layer for purposes of insulation, and he used it to suggest a reason why large aquatic mammals become naked and small ones do not. It was, he argued, because a small animal like a water rat could not produce a fat layer thick enough to conserve the heat to do the job. He wrote:

The thickness of the subcutaneous cellular tissue is in contradiction with the animal's dimensions. The smaller the latter, the relatively thicker should be the fat tissue (a relative magnitude of heat delivery is found to increase in animals of small size). Mammals rather slight in build such as the duck bill or muskshrew, which in principle could develop a subcutaneous fat tissue instead of a pelage fail to do so, because that tissue would be so comparatively thick, much thicker, for example, than in the seal, that it could not be formed in a small animal.[9]

Two other characteristics of fat in aquatic mammals seem to support the AAT hypothesis. One is the distribution of the body fat. In land mammals a high proportion of it is found inside the body cavity, in deposits surrounding internal organs like the kidneys and heart and intestines. In humans, as in many aquatic mammals, it has shown a tendency to migrate outwards to the surface—in some seals, for example, up to 99 per cent of the total is concentrated under the skin.

The second characteristic is the way it relates to the skin. Mammals differ widely in this respect. In an animal like a spaniel, the skin is loose and slides easily over the underlying tissue. The same is true of a rabbit: a cook skinning a rabbit finds that the skin comes away readily in one piece, leaving any deposits of fat adhering to the flesh beneath it. At the opposite extreme is the pachydermous pig, where the fat layer is tightly bonded to the skin. Human fat is at the same end of this spectrum as the pig and the aquatic mammals.

In fact, this particular phenomenon was the seed that originally generated the idea of an aquatic hypothesis in the mind of Alister Hardy. As a marine biologist he had visited the Arctic and often been present at the flensing of whales and other sea mammals, and he was reminded of that when he read the following passage by F. Wood Jones:

The peculiar relationship of the skin to the underlying fascia is a very real distinction, familiar enough to anyone who has repeatedly skinned human subjects and any other member of the Primates. The bed of subcutaneous fat adhering to the skin, so conspicuous in Man, is possibly related to his apparent hair reduction; though it is difficult to see why, if no other factor is involved, there should be such a basal difference between Man and the Chimpanzee.[10]

William Montagna also noted that humans do not resemble other primates in this respect. He speculated that it might be somehow connected with the absence in man of the panniculus carnosus.[11] This is the cutaneous muscle layer found throughout the skin of most other land mammals; it enables a horse, for example, to twitch its skin to try to dislodge flies from areas it cannot reach with its tail. All the primates, including the great apes, have this layer over much of the body. Perhaps our own skin is so thick and adheres so closely to the underlying adipose tissues that the muscles would not be strong enough to make it twitch. For whatever reason, it has vanished except for a vestigial patch in the neck.

There is ample evidence, then, that the difference between human and anthropoid adipose tissue is more than a matter of degree and certainty cannot be attributed simply to the fact that humans are allegedly lazier and eat too much. In attempting to build up the case for the aquatic hypothesis, I felt that the anomalous thick layer of subcutaneous fat was one of the strongest pieces of evidence in its favour. I still believe that—but the claim has by no means gone unchallenged.

Caroline Pond

Caroline Pond, of Britain's Open University, had also begun to think about the problem, which she more than once formulated in her writings: 'Why are humans so fat, and when in their evolutionary history did they become so fat?'[12]

The difference between our approaches was considerable. To me, fat was one of a wide range of topics on which I was reading up, collecting references to what other people had written about it. Pond was out in the front line, doing

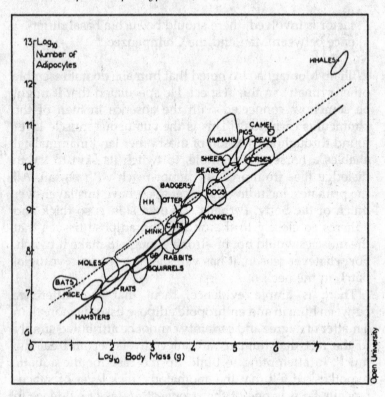

Open University

Comparisons with other mammals show that *Homo* is clearly the odd man out. In proportion to body mass, we have at least 10 times as many fat cells (adipocytes) as expected. The dotted line shows the regression line calculated for 90 specimens of carnivorous mammals. The solid line, that for 101 non-ruminant herbivores. The rings enclose all data points for each species. Carnivorous mammals have more adipocytes in proportion to their body mass than do herbivores. Humans are rivalled only by hedgehogs and fin whales in their deviation from the general trend. (Caroline Pond 'Fat and figures' *New Scientist*, 4 June, 1987, p. 63).

basic hands-on research, and proving that a lot of things other people had written about it were inaccurate.

It was not a fashionable field for a researcher to enter. True, at the present time it is not too difficult to attract grants for research into adipose tissue, but the donors need to believe that the goal of research is to find out how to get rid of the stuff, not to speculate about how and why it got there in the first place. Caroline Pond was a pioneer who approached it in a spirit of pure scientific curiosity. Most of the researchers into adipose tissue at the time confined themselves to analysing small samples of it, usually excised from a deposit of fatty tissue near the testes of the laboratory rat, but she wanted to work on a broader canvas and make inter-species comparisons. Working on a shoe-string, she proceeded to beg, borrow—or rush to the site of accidental death of—cadavers of hundreds of wild, zoo and farm animals, and later Arctic ones, and examine the fat deposits in all of them. She made a number of fascinating discoveries.

One was that fat—formerly regarded as an inert, boring, amorphous substance—had an anatomy and an evolutionary history. It arose in the course of mammalian evolution from eleven distinct depots which can be identified in most living mammals.[13]

The 'subcutaneous' location frequently referred to in previous scientific papers was not one of them; for that reason she makes a habit of confining it within quotation marks. What happens when a mammal becomes very fat is that the adipose tissue from the primal depots spreads outwards until it reaches the surface of the body, and there it may run together to form a continuous layer. A few people wrote to me and advised me that my comments on subcutaneous fat were valueless, since it had now been revealed that there was no such thing.

The discovery about the anatomy of fat was of considerable importance, but it made no difference one way or another to AAT. I was using the term subcutaneous in a purely descriptive sense, referring not to original anatomical nodes from which it had originated, but to the location it presently occupies. *Homo sapiens* is one of the species in which deposits of adipose tissue expanded sufficiently to reach the outer surface of the body and run together to form a continuous layer. The other animals in which this has taken place are hibernating and aquatic ones. That fact may or may not be significant. But its significance is not affected by the words which are used to describe it.

Another highly relevant point made by Pond explains why a human can in some circumstances expand to the girth of a Japanese Sumo wrestler.

Fat does not function in the same way in different species. When a laboratory rat puts on weight its adipocytes (fat cells) increase in size but the number of them does not change. In primates, however (including humans), the adipocytes tend to be smaller, but their number can be increased. That is our bad luck. A fat cell cannot swell to more than three times its original size however much lipid is available to be stored in it. So, if we were not primates and capable of adding to their number, there would be a lower limit to how fat we could get.

It also means that a small percentage of monkeys and apes which are kept in captivity without exercise and with plenty to eat, *do* become fat when they get old, and the excess fat in these specimens begins to distribute itself in roughly the same pattern as it does in humans. So, it is argued, we are not looking at a specifically human tendency for the adipose tissue to desert the internal depots and fly to the periphery: it is merely that 'An apparent shift in the *distribution* of adipose tissue arises as a direct consequence of an increase in its *abundance*.'[14]

But the same thing applies to aquatic mammals like the seals. Accepting that the distribution difference merely reflects the greater abundance of fat, we still have to account for the greater abundance in *Homo* than in the chimpanzee.

The inherited primate capacity to increase the number of adipocytes may have made it *easier* for humans to become the fattest of the primates, but it does not explain why it was adaptive for them to do so, and especially why it was adaptive and obligatory for their babies to do so. The fact remains that in most primates the capacity to pro-

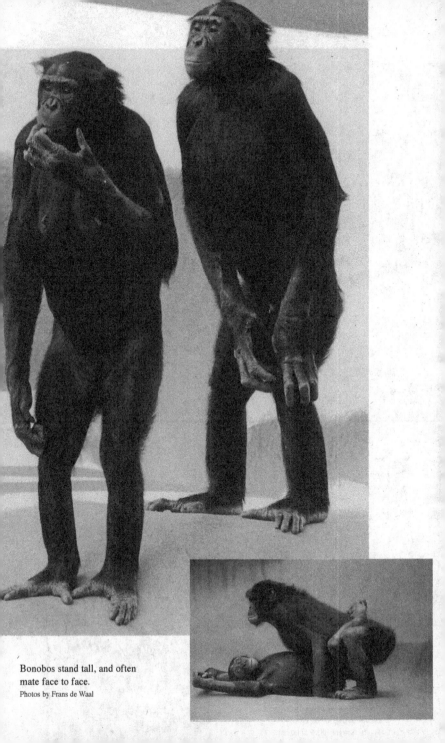

Bonobos stand tall, and often
mate face to face.
Photos by Frans de Waal

'If AAT were true, we would be more streamlined'. More streamlined than what?

Photos: Telegraph Colour Library.

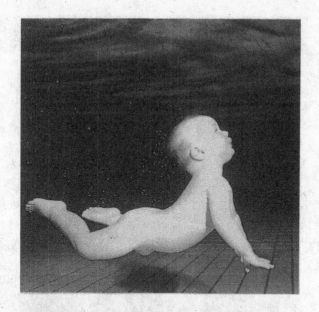

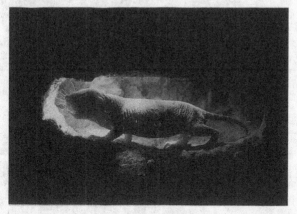

Baby under water, mole rat underground.
Apart from the pachyderms, the only naturally naked land mammals are
Homo sapiens and the naked Somalian mole rat. Photos: baby, Jessica Johnson
and Michael Odent, mole rat © Neil Bromhall/Oxford Scientific Films

The land may vary more;
But wherever the truth may be --
The water comes ashore,
And the people look at the sea.
                                    Robert Frost

duce vast numbers of adipocytes remains latent. Only a tiny minority in very special circumstances become obese. And 'obese' here is a comparative term. An obese monkey is about as fat as an average human being.

One of the controversies Pond engaged in concerned the *raison d'etre* of fat. Its original principal function was as a long-term energy store, as in the hibernators. She queried the widely accepted idea that the main if not exclusive function of superficial adipose tissue was to insulate the body against heat loss, or against mechanical damage.[15]

For example, some new-born seals emerge onto the ice with almost no subcutaneous fat to keep them warm: they become fat by the time it is necessary for them to enter the water. Also, the amount of fat or blubber encasing many Arctic aquatic mammals does not correlate with their need of keeping warm, but often varies within the same latitude according to their feeding habits. Surface feeders tend to have very thick layers of fat which contribute to their buoyancy, whereas for deep divers too much buoyancy would be a handicap, and their fat layers are proportionally thinner. 'Bottom feeding species such as walruses and bearded seals have thick, collagenous skin, relatively little subcutaneous fat, and a massive skeleton, while seals that feed nearer the surface have ... relatively thick blubber'.[16]

I found it hard to understand why she described the insulation hypothesis as 'a major tenet of the Aquatic Ape Theory'. It was, as we have seen, equally a tenet of the savannah theory that the hominids lapped themselves in a coat of fat to keep them warm at night. As for AAT, if buoyancy is included as an auxiliary function of the fat layer, that can only be good news. Water is the only habitat in which it would be relevant.

It is also the only factor which just conceivably makes sense of the fact that the fattest time of our lives is in our infancy. If we consider the mosaic model of hominid evolution—ex-arboreal apes beginning to walk on land but spending time in the trees and sleeping there—then fat would have been a disaster.

In all arboreal primates the commonest cause of infant death is by a young monkey or ape falling from the tree and hitting the ground, especially when it is first learning to move around in the branches on its own. If it were to get fatter, the chance of losing its grip, or the chance of a smaller branch giving way beneath it, would be greater and infant mortality would increase. On the other hand, if it was in an environment where it would fall not onto the ground but into the lake, then the fatter it grew the greater would be its chances of survival. Human infants are born fat and get fatter. A well padded human infant can float on the surface of the water without any support.

I have not seen this recorded in the scientific literature, but I have seen vivid sequences in an Australian film of a baby floating unsupported, resting comfortably on the surface of the water, accompanied by a commentary stating: 'This baby floats happily in water and has done so since birth.'[17]

Sex differences

One other remarkable aspect of human adipose tissue has still to be considered—the difference between the sexes. It has been estimated that 15 per cent of the body mass of a typical young man's body consists of fat; in a typical young woman the figure is 27 per cent. In Western societies, where people tend to get fatter with age, the figures are 28 per cent in older men and 40 per cent in older women.

There is also a well-known difference in the way the fat is distributed, leading to the concept that women's breasts evolved to serve as epigamic markers, indispensable for sexual attraction.

This belief is partly a cultural artefact. Earlier in this century in the West breasts in young women were decreed unfashionable, and the top models anxiously bandaged their chests to preserve the boyish look that was considered sexy at the time. There is no evidence that young men lost their interest in them on that account.

Caroline Pond discounted the idea that women need to store fat in their breasts to provide milk for their children. Most other mammals—from the mouse to the lion—succeed in producing and feeding much larger numbers of offspring without permanent swellings around their nipples. The additional requirements of breast-feeding are normally dealt wih by an additional intake of foods rather than depletion of fat reserves.[18] And if reserves need to be called on, most of the lipid content of human milk is mobilised, not from the breasts, but from the fat deposits in the groin, just as it is in other mammals, whose nipples are not situated on their chests.

One possibility is that the breasts enlarged as a consequence of the loss of body hair and the increased helplessness of human babies when they began to be born at a much earlier stage of development. Most primate mothers sit erect to feed their infants, and most primate babies are able to reach the nipple by clinging to their mother's fur and holding their heads up and their torsos erect. Human babies can do none of these things. A human mother sitting erect to suckle has to deal with an infant lying inert in her lap, unable to lift its weighty head even an inch towards its objective.

Its needs would be better catered for by a pair of nipples lower down—and indeed natural selection has gone some way to answering those needs. The nipples in a human fetus begin to move downwards at the same prenatal stage when the nipples of an ape fetus begin to move upwards.[19] It is possible that having nipples which adhere less tightly to the chest wall, by increasing the mammary lining of fat, was an auxiliary way of making them even more accessible to a supine infant.

Whatever the reason for the sexual differences in the amount and distribution of body fat in *Homo*, it cannot be advanced as a reason why the species as a whole became fatter than other primate species. There is almost no limit to the bizarreness of the epigamic adornments some species invest in, in the attempt to attract the opposite sex, but one rule is absolute: epigamic adornments appear at puberty. No feature which has evolved purely as a sexual attraction is found at its highest peak of development in young infants of both sexes.

It would be possible to imagine an AAT reconstruction of events, based on the perception that human females appear to be better water-adapted than males. Their bodies are simultaneously more hairless and more thickly lined with fat, a combination characteristic of many aquatic mammals. They can survive immersion in cold water for longer, and one athletic sport at which they can outdo males is long-distance swimming. Conceivably they could have been the first to become habituated to the water. In an environment which combined trees and water (a flooded forest or an offshore island dwindling as the sea level rose) the more dominant males would have had first call on the diminished reserves of their traditional food source and would have continued to confine themselves to it. In any society, long-established dominance tends to lead to conservatism. The hungrier females could have been driven to seek for less familiar things to eat and would have found them in the water.

Like William Montagna on primate skin, Caroline Pond has added considerably to our knowledge of all aspects of her subject, and continues to do so, having recently gone on to research the relationship between adipose tissue and the immune system. But I can find nothing in her contributions which is incompatible with AAT.

Like William Montagna, when she looks back at her achievements she finds that the original central question (Why are humans so fat?) remains unanswered. In terms of the conventional scenario it has never been—perhaps can never be—satisfactorily answered. As she commented in connection with the differences in fatness between males and females, these matters remain '. . . poorly understood, and are not readily elucidated from animal studies'.[20]

10

Sweat and Tears

Error is the inevitable by-product of daring.
S. J. Gould[1]

In *The Scars of Evolution* I discussed the problem of why man exudes, through his eyes and his sweat glands, greater quantities of salt water than any other mammal. It is unlikely that these features would have evolved on the savannah where both water and salt are in notably short supply.

I made the suggestion that both the tears and the sweat might have been at one time more concentrated than they are now, and evolved originally for the excretion of salt. I was almost certainly wrong about that, but it may be of interest to recapitulate the course of the debate to clarify the issues involved. It is important to bear one thing in mind: concluding that the evolution of human sweat and tears has nothing to do with salt excretion is not the same thing as concluding that it has nothing to do with water.

PART I: TEARS

The classic AAT question, 'In what other animals has the shedding of non-reflex tears been observed?' produced the beautiful answer, 'In aquatic ones'. And to the further question, 'What purpose do these tears serve?' the scientific literature responded with an unequivocal answer, 'The excretion of salt'.

Since neither the excessive sweat nor the excessive tears

had been satisfactorily explained, I considered the possibility that the emission of these two (chemically not very dissimilar) saline solutions in such unusual quantities might be more than a coincidence. I argued that perhaps they had both at one stage been saltier than they are today, and that part of their original function had been the excretion of excess sodium chloride to restore the salt balance in the body.[2] But the case for this began to unravel, just as the case for the savannah scenario began to unravel, as more facts came to light.

All mammals have some system (lachrymal glands, or Harderian glands) for coating the eyeball with moisture. There is a supply of 'residual tears' just sufficient to keep the surface of the eye moist. Secondly, when the eyes are irritated by noxious vapours, foreign bodies or icy winds, the output of moisture increases and overflows: these are reflex tears. But humans also weep tears which have nothing to do with local irritation of the eyeball. Darwin called them, rather picturesquely, 'psychic' tears. A more neutral term would be simply 'non-reflex' or 'emotional' or 'non-irritant', and it seemed to me that they were not without analogues in the animal world.

In 1959 Kurt Schmidt-Nielsen and his co-workers solved the long-standing problem of how oceanic birds can live without fresh water.[3] These birds have paired excretory glands—situated in grooves in the skull over the eyes—which enable them to drink salt water without ill effects. These glands remove salt from the birds' blood far more efficiently than their kidneys do. It is excreted through a duct in the nasal cavity and shaken off from the bill in drops of brine.

Smaller but similar glands can be found in non-marine birds; they excrete excess potassium rather than excess sodium, and possibly represent the more primitive form of the organ.[4] However, the glands are generally referred to as 'salt glands'. Their existence and function were first discovered in marine birds, and their size and activity in different species vary directly with the percentage of time spent out of range of land and accessible fresh water.

Since our own tears are shed from the eyes rather than the nose, the nasal glands of the sea birds do not seem to offer a very close parallel. But Schmidt-Nielsen in the same paper revealed that there are similar salt-excreting organs in marine crocodiles, marine iguanas, sea snakes and marine turtles. In all these cases the brine is excreted from the eyes in the form of salt tears. If the creatures are induced to ingest salt water, the tears are activated.

Another point stressed in that paper was that this mechanism appears to have evolved independently more than once in different lineages. The location of the glands in turtles, for example, is different from the location in birds and indicates a different evolutionary origin, and in the marine iguana it is a different one again. But the composition of the tears in all of them is strikingly similar.

It seemed possible then that it might have arisen independently yet again in marine mammals. It was not difficult to find references in the literature to the shedding of tears by aquatic mammals such as seals and sea otters, and that archetypal pachyderm the elephant. It was hard to resist the conclusion that our tears might have had a similar origin during an ancestral sojourn in the highly saline Sea of Afar before it finally evaporated, and that later in our evolution, when salt excretion was no longer necessary, the tears (and the sweat) became more dilute, as they are today.

In favour of the idea was the fact that human emotional tears are still in part excretory. William Frey[5] has established that they help to eliminate substances such as prolactin, ACTH, Leucine-enkephaline and manganese. It seemed within the bounds of possibility that if at one time it had been desirable to excrete sodium chloride, non-reflex tears could have served as an auxiliary channel for this purpose.

It is true that there were also recorded references to weeping among land animals—heart-broken dogs, doleful koalas and, according to Dian Fossey,[6] one gorilla called

Coco which shed tears on a single occasion—'. . . something I have never seen a gorilla do before or since'. I (naturally) concluded that these were all anecdotal or misdiagnosed. Coco, for example, was very ill at the time, and even a cat's eyes may water if it has 'flu. The eyes of aged dogs, like those of aged people, are liable to get rheumy, and the orbs of a faithful old spaniel may well seem to be swimming with emotion even though the tears do not flow. The one-time belief in weeping 'bears' can be traced back to the statement of a journalist in *The Illustrated London News*, 21 February, 1931, that when molested the koala 'cries piteously like a child; tears roll down its face, and it rubs its eyes with its forepaws'. But the naturalist on whose account the article was based later demurred: 'It is only a cry of fear and not of pain and I have noticed no tears'.[7]

William Frey, the author of *Crying: The Mystery of Tears*, received a copious correspondence from the owners of weeping pets. His conclusion was:

As I opened letter after letter I began to feel that I must be one of the few persons who had never seen animals shed emotional tears . . . I have asked several owners to provide further evidence of animals shedding emotional tears, either by videotaping the episode or obtaining a testimony from a veterinarian or other animal expert. To date, I have received none. When I asked several small-animal veterinarians if they had ever observed a dog shedding emotional tears, they all told me they had never witnessed pyschogenic tearing in any animal, and added that they felt all animals' tears were due to eye irritation or inflammation.[8]

No hard evidence

But, alas, what is sauce for the goose is sauce for the gander. The references to weeping aquatic mammals, when subjected to equally rigorous scrutiny, showed the same tendency to evaporate. Accounts of travellers interviewed by journalists, and translations from foreign languages, are both rendered unsafe by the fact that—at least, in English—'crying' is a term applied indiscriminately to a week-old infant crying from hunger and a woman crying from grief, though the yelling of the young baby is not accompanied by weeping and the tears of the woman may be silent. I had cherished a quotation by W. Steller, discoverer of Steller's sea cow (now extinct), who described in the 1740s how sea otters deprived of their young would—according to an early translation—'weep over their affliction just like human beings'. But a new translation (from Stanford University Press in 1988) gave a different rendering: 'When the young are taken from them they cry aloud like a little child, and grieve . . .'[9]

Elephants and seals are the species most often described in the literature as given to weeping. Darwin described the elephant's tears,[10] and apart from references in books and the personal testimony of zoo keepers, I have seen the phenomenon on television in an animal forming part of a procession. There are plenty of photographs of seals, especially young ones whose fur is dry and fluffy, with their cheeks wet with tears.

But in both cases the evidence is of a kind that can be discounted as 'anecdotal'. No one has quantified the frequency with which elephants shed their tears or tried to relate the event to any particular factor—whether environmental factors such as temperature, physical such as the salt balance, or emotional such as loss of an infant or separation from the herd or from a keeper.

Similarly with seals, it is pointed out that the seal's tears flow more readily because they lack the nasolachrymal duct which, in humans, channels part of the surplus moisture into the nose rather than down the cheeks. (That is why people trying to hide or check their emotional tears often have to sniff or blow their noses instead of dabbing their eyes.) But that does not account for all the data of the phenomenon of the seal's tears. If it was merely an overflow of the background tear flow necessary to keep the eyeballs moist, the tears would be minimal and found in all the animals all the time. But on a crowded beach only a few at any time seem to be shedding tears, and again there has been no attempt to determine whether this is random or has any behavioural connotation. One comment was that seals weep in hot weather. 'Pinnipeds may also weep copious tears on hot days, though this can hardly cool them much'.[11] But again this factor would be operative in most or all of them at any given time. This class of evidence, then, remains nebulous.

However, I came across one piece of hard scientific evidence which seriously undermined the case for the excretory theory of tears. Apparently, though weeping in marine birds and reptiles is clearly an adaptation for regulating salt balance, this does not apply to marine mammals. Marine birds and reptiles, when induced to ingest salt, duly discharge it from their bills or from their eyes as described by Schmidt-Nielsen. But in his account in *Scientific American* he added: 'We have undertaken, however, some pilot studies on seals. When we injected them with salt solutions that stimulate the salt glands of birds and reptiles, they merely increased their output of urine'.[12]

It would appear, then, that a seal's kidneys, unlike those of birds and reptiles, are quite adequate for the task of controlling the body's salt balance unaided. But then, so are the kidneys of apes and humans. Even in a marine environment the hominids should not have needed to evolve an auxiliary mode of eliminating salt.

Clearly the seal did not fit into my scenario. Even the marine iguana did not fit into it as neatly as I had hoped.

From the scientific papers I read, I had imagined doleful tears stealing down its cheeks, rather like ours, when it felt queasy from its salt imbalance. Then one day I saw a television programme ('In the Wild: The Galapagos Mystery') showing one of these reptiles on the Galapagos, finding itself suddenly eyeball to eyeball with the actor Richard Dreyfuss. In a reaction as explosive as a sneeze, a shower of salt water was ejected from its eyes and sprayed in all directions by violent shakes of its head. This put it firmly in the same camp as the marine birds which use the same gesture. And it went some way to confirming my contention that emotional triggers can get mixed up with salt excretion. But somehow the whole incident left me with the gut feeling that this was not weeping as we know it.

On balance, then, I am inclined to concede that the excretory hypothesis of tears was misconceived. I am not quite certain, because there is still one little nugget of fact which does not fit into any other interpretation. That is the cricopharyngeal spasm—the 'lump in the throat'—which often precedes or accompanies human weeping. The only animal analogue for that seems to be the muscular constriction which the seal applies to a fish passing down its gullet to minimise the amount of salt water that goes down with it.

Despite the lump in the throat, the original AAT theory of tears has flaws in it. If there were any other theory available that had no flaws in it, that would settle the matter.

Alternative theories

Darwin tried hard to find a reason why we differ from the apes in this respect. He focused on the greater variety and complexity of muscles which make human faces so much more richly expressive than animal ones. He suggested that '... whenever the muscles round the eyes are strongly and involuntarily contracted in order to compress the blood vessels and thus to protect the eyes...' (during, for example, a screaming fit in a young child) 'tears are secreted, often in sufficient abundance to roll down the cheeks'.[13] In support of this, he noticed that when the orbicular muscles are contracted for other reasons, such as violent coughing or vomiting, tears are often shed.

But tears are often shed silently and with no muscular contractions. And Darwin himself noted that young infants may scream violently with their eyes tightly closed, yet no tears are shed because the lachrymal glands have not 'come to full functional activity'. However, that last statement needs qualifying. It is only the shedding of *non-reflex* tearing that has not begun to function. When he accidentally touched the eyeball of one of his infant children with the sleeve of his coat, the reflex tears flowed quite freely.[14] He offered no reason why the muscular contractions around the eyes did not cause the shedding of tears as readily as the touch of a coat sleeve.

Ashley Montagu in 1959 proposed that the shedding of tears was a protection against damage to the mucous membranes of the nose caused by 'dry crying', because screaming and sobbing involve excessive intake and expulsion of the breath which dries up the membranes.[15] He argued that in early man 'those individuals were naturally selected in the struggle for existence who were able to produce an abundant flow of tears since the tears acted as a preventative of mucosal dehydration', and he explains the differences between men and apes by referring to the longer dependency of human infants.

The weakness of this theory is that while young apes do not remain physically as helpless as human babies over such a long period, they do continue to indulge in prolonged noisy tantrums and what he calls 'tearless crying' right up to the time they are weaned. Also the deleterious effects of dry crying which he lists affect the nasal passages alone. Screaming and breathing in and out do not dehydrate the eyeballs. To cope with these problems natural selection would have favoured insulatory secretions in the nasal passages themselves—as in colds and catarrh—rather than the shedding of saline fluid from the lachrymal glands, most of which flows uselessly away.

E. Treacher Collins (1932) suggested that weeping 'has arisen in human infants as a purposive action to attract attention with the object of eliciting aid and sympathy'.[16] It may often be resorted to with that object, but it can hardly have evolved for that purpose. A human baby, like its anthropoid cousins, is very good at attracting attention. It can open its mouth into a wide square-shaped hole, distort its features, go red in the face, and utter ear-splitting and heart-rending noises. Any primate mother who could remain oblivious to all that would not have her attention riveted by the addition of a few drops of colourless liquid.

Then there is always the last resort argument—that there is nothing to explain, that all animals have a lachrymal apparatus and are capable of shedding tears, and some humans just happen to shed quantitatively more.

The difference is not merely quantitative. Emotional tears are different in content, are secreted in response to different stimuli, and are differently innervated. If the eye is irritated by smoke, grit or noxious vapours the response is instant and unconditional. A message travels inward along the fifth cranial nerve to the lachrymal nucleus in the brain stem to say in effect, 'Ouch!'. And a message instantly travels back along the seventh cranial nerve to stimulate an immediate increase in tear output. There are certain physical disorders (such as the condition known as 'dry eye') in which the production of background tears and reflex tears is impaired, yet the emotional tears flow as copiously as ever. They are controlled by a different nerve, which travels down from the frontal cortex to the brain stem. As William Blake expressed it: 'A tear is an intellectual thing'. In this respect, as in many others, *Homo sapiens* appears to be unique, and no one has convincingly explained why.

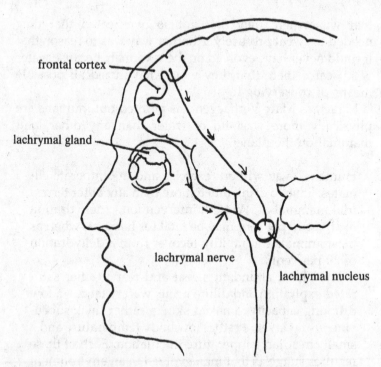

frontal cortex

lachrymal gland

lachrymal nerve

lachrymal nucleus

When smoke gets in your eyes, the message is carried to the lachrymal nucleus via the lachrymal nerve. If sad films make you cry, that requires some input from the frontal cortex before tears will be shed.

PART II: SWEAT

If you visit the Bronx Zoo in midsummer you may notice a curious phenomenon. The skin of the human animals outside the cages is glistening, if not actually streaming, with sweat. The skin of all the animals inside the cages appears to be dry. As William Montagna observed, 'The dissipation of heat is the function that most conspicuously distinguishes human skin from that of other mammals.'[17]

This has been interpreted by evolutionists in the past in one of two ways. One way was to reason that our ancestors must have evolved this characteristic on the savannah—

that was where they first got so overheated that they needed to sweat profusely. Another way was to reason that it could never have evolved on the savannah, because water is so scarce there that they would have used all possible means of conserving it.

In fact, as Marc Verhaegen has pointed out, humans are physically more wasteful of water than any other land mammal on the planet.[18]

> Humans, even without exercise and in temperate climates, have to drink much more than any other terrestrial mammal . . . Without intervention, a dehydration of about 10 per cent may be fatal for humans, whereas most animals can rapidly recover from a dehydration of 20 per cent . . .
>
> They have abundant sweat and tears, rather saturated expiration and dilute urine, watery faeces, a low drinking capacity, a naked skin, a rather thick subcutaneous fat layer, a rather low body temperature and a small circadian temperature fluctuation. Each of these features suggests that man evolved in an environment where water was permanently and abundantly available.

One of the oldest and most universal mammalian methods of dealing with overheating is panting. It is used—either as the main or as an auxiliary method of thermoregulation—by virtually all other land mammals including the African apes. But some time after the ape/human split the hominids dispensed with it. Modern humans pant during exercise to increase their oxygen supply, but have lost the trick of doing it to reduce their temperature on a hot day.

Peter Hiley reported of panting: 'This response to heat exposure has been found to occur in all primates investigated to date, except in man . . . The questions then arise of how and when during his evolution did man lose the ability to pant. The answers to these questions have yet to be found'.[19] It has even been suggested that panting disappeared following the evolution of speech because it would be hard to pant and talk at the same time.[20]

The eccrine gland mammal

Instead of panting, humans rely on sweat glands to keep them cool. But they do not use the same kind of sweat glands as, for example, cattle and horses. *Homo sapiens* has been described as the 'eccrine gland' mammal—because we use eccrine glands for sweating.

Eccrines are found in most mammalian species, where they are confined to the pads of the feet, as in dogs and cats. They exude a saline liquid, and their function seems to be to prevent slipping. Hoofed animals—which do not have pads on their feet—have no eccrine glands.

In lemurs the eccrines are still confined to the volar surfaces (palms of the hands and soles of the feet), and that remains true of all prosimians and all of the New World primates except for four species.

These four are the spider monkey, the woolly monkey, the howler monkey and the woolly spider monkey. They are the only species with truly prehensile tails which can grip on to a branch and support their weight. In these four species the under-surface of the tail is richly supplied with eccrine glands as if it were a third hand, which in some functional ways it resembles. It would seem that having once escaped the 'palms and soles only' restriction, the eccrine glands felt themselves enfranchised to crop up wherever they liked, because they are found randomly scattered all over the skin of these four species. The random scattering is also found in all the Old World apes and monkeys whatever their habitat, suggesting that they may all have been descended from one or more common ancestors which once had prehensile tails. In most cases these ectopic (displaced) glands appear to serve no obvious purpose.

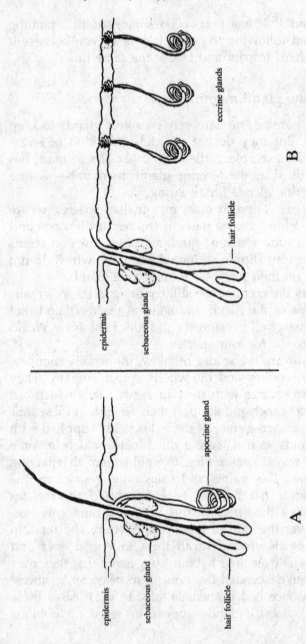

Skin glands. In most mammals (A), over the general body surface, all the skin glands are connected to the hair follicles. Eccrine glands are found only on the volar surfaces (soles of the feet, palms of the hands). In humans (B), the hairs are attenuated, the apocrine glands disappear before birth, and millions of eccrine glands exude sweat directly onto the skin.

In the gorilla, for example, the eccrine glands are quite numerous (they have about half as many as we do), but they do not use them for keeping cool. William Montagna found that very puzzling. 'Why,' he queried, 'do these glands not function when they seem to have all the equipment for doing so?'[21]

At some point between the apes and humans, the eccrine glands in humans became twice as numerous and went into action as the primary mode of temperature control.

There has always been a general predisposition to assume that everything unique about *Homo sapiens* must be a great improvement on anything evolved by any other primate. Despite this, several of the more independent-minded scientists took a dim view of eccrine sweating as a means of keeping cool. Russell Newman pointed out that when the temperature goes up, the eccrine glands are slow to respond and take at least 20 minutes to reach peak production.[22]

When it does get under way, it is unnecessarily and wastefully profuse as compared with the apocrine sweating of an animal like a sheep. William Montagna concluded that it was not only 'an evolutionary blunder', but also that the human eccrine glands must have originally multiplied for some other purpose, and only later been conscripted into the business of thermoregulation.

There had to be some way of making sense of all these anomalies. My first approach was to focus on Montagna's suggestion that the non-volar eccrines began to multiply for some purpose other than sweat cooling. Influenced by my tears hypothesis, I mooted the idea that the eccrine glands may also at one time have served the purpose of salt-excretion. Two factors seem to sustain the idea:

1 Eccrine sweat—unlike that of the apocrine sweat-cooling glands used by non-primates (which tends to be oily)— is composed almost exclusively of water and salt.

2 Although in modern humans it is normally too dilute to

correct any overconcentration of salt in the blood, it can still apparently fulfil this residuary function if the need arises. Ingram and Mount observed: 'Sweating can be regarded as an excretory process that takes over from the kidneys during heat stress, when the volume of urine is reduced'.[23]

The excretion theory was speculative and controversial, but there seemed little danger that the factual basis for this speculation would begin to unravel. It was not, like the tears thesis, based on 'anecdotal' evidence. It was all drawn from respectable academic sources.

Nevertheless, unravel it did. Researchers began peering very much more closely at the other primates. Traces of 'cutaneous moisture' were detected on the skin of the baboon.[24] Also, the rhesus monkey.[25] And the macaque.[26] And the chimpanzee.[27] I received a (doubtless subjective) impression that a large number of primate species which had remained cool and dry throughout the first half of the twentieth century had suddenly begun sweating like stevedores.

There was, understandably, a tendency to concentrate on savannah-dwelling species (with the curious exception of vervet monkeys. The last we heard of them was that they do not sweat,[28] but no doubt somebody will catch up with them and make them do it).

In 1980 a classic study of the patas monkey was published by Sheila A. Mahoney,[29] establishing that the patas monkey sweats more profusely than any primate except man and that this is very effective in controlling its temperature. For some time I clung to a last-ditch belief that it might have been apocrine sweating; all non-human primates have apocrine glands all over their bodies. I read her paper carefully but this was not specified. She was more interested in measuring the amount of moisture produced than establishing where it came from.

Then, in 1988, Reynaldo S. Elizondo laid it on the line

with a paper entitled: 'Primate models to study eccrine sweating'.[30] There was no longer any legitimate room for doubt: the little patas monkey, the second sweatiest primate and the fastest running primate on the savannah, uses eccrine glands for sweat-cooling, just as *Homo sapiens* does.

The New Scientist once published a cartoon of a laboratory in which a scientist had been examining the contents of a test-tube. He looked up glumly and enquired of his colleagues, 'What is the opposite of "Eureka"?' I know just how he felt. I was not about to suggest that the ancestors of the patas monkey had passed through an aquatic period. And I was not about to propose that eccrine sweating had evolved in the patas monkey for one reason, and in humans for quite another. I had too often criticised opponents of

Patas monkey. *Amanda Williams*

AAT for taking that line in connection with nakedness and fat.

The alternative explanations

However, the patas monkey put a spanner in the works not only for me but also for most evolutionary theorists. It had frequently been argued that human nakedness was a direct consequence of eccrine thermoregulation, that the sweating would not have worked without it. Nakedness was said to be 'adaptive . . . to hot dry conditions in which the surface of the skin *must be free* to permit the breezes to evaporate sweat' [my italics].[31] Peter Wheeler argued that body hair 'prevents the efficient use of sweating for dissipating heat'.[32]

There had always been a dissident minority, like Russell Newman[33] ('. . . no evidence that a hair coat interferes with the evaporation of sweat') and V. Sokolov[34] ('. . . fur does not prevent the evaporation of sweat') and the patas seems to have proved this point. Its coat is quite thick and red and shaggy, and yet however great its exertions and however hot the sun, it does not collapse from hyperthermia in the daytime nor have to shiver in the tropical night.

The question now is how many extant theories fit all the facts as we know them today.

Peter Wheeler's first thesis had suggested that three crucial changes—bipedalism plus sweating plus nakedness—were closely interconnected and that all of them occurred following the move to the savannah. It is now clear that the bipedalism evolved very early and in a habitat where hyperthermia would not have posed a major problem. Up to a point he has conceded this by softening the savannah backdrop ('In this connection the term "savannah" does not necessarily mean open grassland')[35] and calculating that an intake of water 'as low as 0.7 litres may have been sufficient to replace all losses incurred by a naked biped throughout the day if it retreated into the shade for a 4-hour period in the afternoon'.[36]

So the original tightly-knit trio of alleged thermoregulatory adaptations has come apart. The bipedalism arrived before the move to open ground; the open habitat was a tolerable one, allowing for four-hour siestas in the shade and not subjecting the hominids to intolerable pressures demanding unique solutions; and the nakedness was neither an inevitable consequence nor an essential precondition of the profuse eccrine sweating (*vide* the patas monkey).

Russell Newman's thesis is not undermined by the patas monkey, because he never attributed nakedness to savannah conditions. He believed that the prehominids were already naked as well as bipedal before they left the shelter of the ancient forest habitat. The weakness of Newman's version is that he offers no reason why, among the apes that shared the ancient forest habitat, only one group became naked. With sweating, as with tears, the short-lived excretory theory was not superseded by a more satisfactory one. It was superseded by the question mark that had been there before it arrived.

PART III: THE CONSEQUENCES

It is sometimes implied that retracting this theory seriously undermines the standing of the aquatic hypothesis. The credibility of AAH would indeed be weakened if the salt excretion theory had been the major reason for believing in it; or if it had not been based on scientific data that had appeared valid at the time; or if I had continued to believe in it when the balance of the accumulating evidence clearly swung against it; or if I had continued to defend it in public after I had privately ceased to believe it; or if there was an emerging consensus in support of some other explanation of why *Homo sapiens* is the leakiest primate.

None of these conditions obtains. So the consequence of

retraction is that once again we are back to Square One.

The best strategy at Square One is to begin asking simple basic questions. In connection with tears, the first question might be: What are tears for? Apart from lubricating the eyeball and, when necessary, helping to flush out foreign bodies, they contain—like the mucous lining of the nasal passages—some bacterial properties to combat infections.

Question 2: Is there anything about the eyes or the tears of primates which distinguishes them from the eyes and tears of other orders of mammals?

Most primates lack the auxiliary eyelid known as the nictating (or nictitating) membrane.[37] This organ dates back to before the reptile/mammal split. In aquatic reptiles like crocodiles, this third eyelid is a transparent sheet that moves sideways across the eye from the inner corner, cleansing and moistening the cornea without shutting out the light. In mammals that possess it, its operation usually coincides with a blink, so that among domestic pets the only one in which it is often visible is the parrot.

Question 3: Does the lack of this additional layer of protection in primates mean that the bactericidal property of their tears needs to be enhanced?

Apparently it does. The tears of the higher primates contain an active substance called lysozyme. Alexander Fleming was the first to demonstrate that lysozyme is highly bactericidal.[38] It has been shown to inactivate the viruses of many infections such as herpes simplex, herpes zoster, warts, vaccinia, and so on.[39]

In the apes the lysozyme is contained in the 'residual' supply of lachrymal fluid which most mammals secrete. For this purpose the production of tears has to be constant—but it is strictly minimal, merely sufficient to coat the eyeball.

Question 4: Is it possible to think of circumstances in which that minimal supply might be insufficient? Conceivably, yes. If the eyes were employed from time to time for looking through water instead of through air—and, unlike the crocodile's, had no third eyelid to protect them—then the tears secreted and the bactericides they contained would be washed away as rapidly as they were produced, necessitating the production of a more copious supply of lachrymose fluid and possibly a higher concentration of protective chemicals.

Question 5: We know that *Homo*'s supply of tears is frequently more copious. Is there any reason to suppose it is more bactericidal?

V. M. W. Bodelier and colleagues in 1993 researched the differences in content between chimpanzee and human tears and reported:

> In human tear fluid, relatively high concentrations of proteins like immunoglobin A, lactoferrin, tear specific prealbumin, and lysozyme are found, beside high activity of the tear specific enzymes peroxidase and amylase ... The species comparisons performed until now indicate that human tears remain unique in the high content of lactoferrin.[40]

All of these data may be coincidental. There may be other explanations for them which have not been considered. They are included to stress the point that the question remains unresolved, that it may require some lateral thinking to find the solution and that an aquatic connection offers a possible pathway regardless of whether the water was saline or fresh.

The connection between water and sweat is plainer. It accepts the Newman hypothesis that the hominids became hairless in the same habitat where they became bipedal— in the forest. It adds the proposition that, following some kind of ecogeological event, some descendants of the l.c.a. continued living in arboreal habitats with a dry forest floor, while others found themselves in a combination of forest and water—possibly flooded forest, or mangroves, or an island with diminishing tree cover.

It was probably during this period that they lost the

capacity to pant. Panting, it has been argued, is the most efficient way of dissipating heat which flows in from the environment (as opposed to being generated by exercise). If they were not merely living in the shade but also spending much of the daytime in water, the need of dissipating environmental heat would have been minimal or absent.

A watery habitat is by far the commonest explanation of loss of body hair in mammals. It would be unparsimonious to suppose that this would not also apply to the nakedness of the naked ape.

Very much later, when their descendants began venturing away from the trees and the waterways and spending some time on the open plain, they would already have been naked bipeds. That was what made them so different from other primates (the vervet, the patas, the baboon) that moved to the savannah. It means that the causal connection between the sweating and the nakedness was the reverse of the one which has been commonly canvassed. They did not have to become naked because they were sweating. They had to sweat profusely—even more profusely than the patas—because they were naked under a hot sun.

11

The Larynx and Speech

It seems that within the evolution of the human larynx may lie many of the secrets to understanding how we came to be.

J. T. Laitman and J. S. Reidenberg[1]

The descended larynx is another of the anomalous characteristics of *Homo sapiens* that distinguishes us from all the other primates.

It has inspired a lot of evolutionary speculation because many people believed—many people still believe—that this was the feature which enabled us to learn to speak. The power of speech separates us from the rest of the animal kingdom more profoundly than all the other anomalies put together.

When we pose the classic AAT question, 'In what other species can this feature be found?' the number of known species is disappointingly small. However, there is no mistaking what they have in common. A descended larynx is found in the dugong and the walrus and the sealion. (There may be others which have not been identified.) Of the sealion, for example, V. E. Negus wrote: 'There is practically no epiglottis, and the larynx cannot in consequence take up an intranarial position.'[2] Not all aquatic mammals display this feature: in the whale the larynx has moved up rather than down. But it may possibly be significant that this feature has up to now been found in no terrestrial mammal other than *Homo sapiens*.

What 'descended' means

In a TV sitcom about a group of aliens visiting the earth, one of them discovered with alarm on first finding himself in a human body: 'I have three holes in my face!' It is, of course, a standard condition for a mammal to have three holes in its face. Two of them (the nostrils) lead to a channel which passes down through the windpipe (the trachea) into the lungs. This channel makes possible two major functions—breathing air and collecting information about the environment by means of the sense of smell. Another channel leads from the mouth via the gullet (oesophagus) into the stomach, and serves the purposes of eating and drinking.

In most mammals these two channels are kept strictly separate all the way down by an ingenious arrangement which has been called the two-tube system.

The pristine and perfect form of the two-tube system is found in reptiles. The upper end of the windpipe passes up behind the hard palate (the roof of the mouth) into a space at the back of the nasal passages. It is locked in there, ensuring that all air that is inhaled passes without let or hindrance down into the lungs. Mouth breathing is imposs-

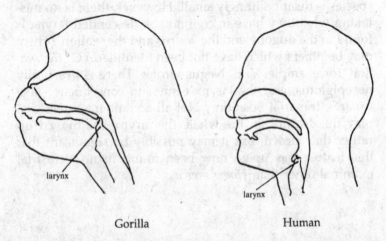

larynx

larynx

Gorilla   Human

ible. Anything, liquid or solid, taken in by mouth has to pass to either side of the windpipe on its way to the stomach. Thus, no air can get into the stomach, and no food or drink can ever get into the lungs. The system is foolproof.

The arrangement in mammals is a slight modification of this, possibly in the interest of vocal utterance. (Most reptiles are dumb.) At the upper end of the windpipe is a structure called the larynx, which contains the voice box. In mammals this is not locked into the back of the nasal passages; it can move up and down. A dog, for example, can retract the larynx downwards in order to bark or to pant. When it has finished barking or panting the larynx goes back up like a periscope into its resting-place above the palate, and nose breathing is resumed.

Even when the dog is panting, although its mouth is open, it is not breathing *in and out* through its mouth.[3] It is breathing in through its nostrils and out through its mouth. One reason may be that it cannot afford to miss any incoming scent signals. Keeping the olfactory information on stream is as important to a dog as keeping our eyes open is to us.

In this model the larynx remains above the palate except during three functions—panting, making vocal utterances, and sometimes swallowing. If the animal eats a lump of food which cannot pass easily around, or to one side of, the windpipe, the larynx is drawn down out of the way to allow it to pass.

This modified two-tube system is the standard pattern among mammals and is very nearly foolproof. It is the system used by all non-human primates. It is also used by human babies. When suckling they can drink and breathe at the same time with no danger of the milk 'going down the wrong way'. But the larynx begins to sink lower as the baby grows, and in an adult human it has ended up right down in the back of the throat, where the two tubes—one into the lungs and one into the stomach—lie open side by side.

Disadvantages

Charles Darwin thought it was a crazy arrangement. He pointed out that '... every particle of food and drink we swallow has to pass over the orifice of the trachea, with some risk of falling into the lungs'.[4] Choking on inhaled food is not a common cause of death, but the sudden collapse of a customer in a restaurant—with symptoms misleadingly reminiscent of a heart attack—is frequent enough for some casualty departments to have christened these cases 'cafe coronaries'. Similarly, other people die through inhaling their own vomit—another exclusively human hazard. And because of the lack of an uninterrupted airway from nostrils to lungs, unconscious traffic accident victims may die unless someone quickly applies the First Aid maxim: 'First check the airway'.

Much more prevalent than the fatal incidents are the disorders which impair the quality of life as, for example, the chronic bronchial disorders and throat infections commoner in our species than any other. Breathing through the nose has evolved in mammals over millions of years. It ensures that dust particles are filtered out of the inhaled air by hairs lining the nasal passages, and that the air is warmed and moistened and partially sterilised by secretions of the mucous membranes before it reaches the lungs.

Mouth breathing means that all these functions have to be dealt with by the vulnerable lungs, which were never designed to cope with them. Humans are all mouth breathers for part of the time, when taking exercise and talking. Untreated respiratory infections in childhood may lead to *habitual* mouth breathing, with increased liability to infections of the throat and the lungs and sometimes of the inner ear.[5]

Later in life, with loss of muscle tone, another consequence of a descended larynx is a liability to incur obstructed sleep apnoea—repeated blockage of the airway during sleep. The sleeper may not be awakened to full consciousness by these recurring episodes, but the deprivation of REM (Rapid Eye Movement) sleep can lead to anxiety, depression, and chronic fatigue during waking hours.[6] Recently a lorry driver was cleared of guilt for a traffic accident on the grounds that he suffered from this condition.

Finally, there is the connection between the descended larynx and cot deaths or SIDS (Sudden Infant Death Syndrome). The discovery of this is fairly recent. A paper published by J. J. McKenna in 1986 gives a valuable background to the nature of the problem and the way it was being discussed at that time.[7]

One point made very clearly is that the problem of SIDS is 'unique among mammals'. 'SIDS does not occur in other species.' Another is that it occurs only during the period when the larynx is beginning to move down to the adult position and conscious breathing is beginning to replace reflex breathing: 'No other infant malady except for infant botulism is so consistently and narrowly delineated by age . . .' 'No hypothesis is viable unless it can explain the syndrome's restricted age distribution.'

It would seem likely, then, that either laryngeal descent or conscious breath control might have some relevance to the condition. McKenna laid most stress on breath control. J. T. Laitman, in the discussion appended to the paper, commented that 'true' laryngeal descent does not occur until about 18 months. But Laitman adds: 'We have noticed that the first instances of oral tidal respiration are found in infants between 4 and 6 months'. The salient factor, then, seems to be not when the larynx is fully descended (by that time the danger of SIDS is over) but when it first loses its secure contact with the palate.

Edmund Crelin published a paper on it,[8] suggesting that

the difficulty arises when the infant's larynx is no longer securely locked in above the palate, and not yet safely tucked away below the base of the tongue, but at the back of the mouth on its way down. He thought that at this vulnerable time, when the baby was lying prone, there was a possibility that the uvula could enter the opening of the windpipe and block it up, causing asphyxia and death. After death, when the baby was picked up and moved, the larynx would lose contact with the uvula and the autopsy would reveal nothing amiss.

He suggested that the danger would be lessened by changing the infant's posture—possibly by lifting its head a little. Some lives were saved by this practice in a clinic where it was applied. But the really big life-saver came when this advice was amended to a warning by baby clinics against letting babies sleep in the prone position, that is, face down. (The warning was important because in some countries mothers had been regularly advised to lay their babies face down, in the belief that it would lessen the chances of their choking on their own vomit.)

The advice to lay the babies on their backs was promoted in Holland in a nation-wide campaign, and it cut cot deaths by 40 per cent in one year.[9] In 1991 Britain launched a 'Back to Sleep' campaign (that is, 'on their backs to sleep') and in 1994 a newspaper headline proclaimed: 'Cot deaths show 70 per cent drop over past 5 years'.

There may be other factors involved, such as general health or infections, but the scale of success of the 'Back to Sleep' advice seems to provide overwhelming evidence that Crelin got the priorities right. The only thing changed by following this advice is the direction of the force of gravity relative to the baby's respiratory organs, acting on the only organ loose and mobile enough to be affected by it. That is the unattached upper end of the infant's larynx.

Negus and the larynx

It is clear, then, that the descended larynx in man is a very rare feature, attended by many disadvantages and no obvious advantages. Earlier this century a man called Victor Negus addressed himself to the question of how and why it had evolved. He spent the rest of his life dissecting, analysing, drawing and describing the larynx in every species he could lay his hands on. His output of books and papers has been described as 'prodigious'.

He was never tempted to believe that the larynx had descended in order that the hominids would be able to speak. Evolution takes place in response to something that has already happened or is happening, not because of something that may happen at a future date. He stated this belief more than once: 'To think that the larynx and its vocal cords have been evolved primarily for the purposes of speech is incorrect'.[10] He believed that '... much, if not all, of the anatomic mechanism was evolved primarily for purposes other than speech'. It only remained to discover what those purposes were.

He began work in the 1920s, when there was a general belief that everything that was different and distinctive about the human race was axiomatically an improvement on anything that had preceded it.

But the more closely he studied the gap that was interposed between the epiglottis and the palate, the less he found to celebrate. He became keenly critical. 'Today,' he wrote in *The Mechanism of the Larynx*, 'man can breathe through the mouth, to his great advantage.'[11] Later in life, in a condensed version of the book, he opted for a mellower version: '... to his partial disadvantage'. But he repeated his conviction that the transition to mouth breathing '... leads to many evil effects'.[12]

After 20 years of study of the subject he had adopted an approach in which he spoke of the larynx being '... forced to descend', rather than embracing the opportunity to descend. The big question was: What forced it? In the minds of most commentators that remains the big question.

What forced it down?

The most simplistic suggestion is that gravity alone pulls it down when the child begins to stand up and walk. That was questioned by R. van den Berg and Jan Wind, and their researches established that even in adult patients who have for some reason always remained supine, the larynx still descends.[13]

Another suggestion connects it with the angle of the head. J. T. Laitman has found a correlation between the position of the larynx and the orientation of the basi-cranium.[14] But the correlation does not necessarily indicate that a change in the degree of flexion *caused* the descent of the larynx—only that these two conditions coexist in mature humans and that both are absent in all other mammals.

One popular theory is that it happened because our faces are flatter than those of apes, and when the jaws receded, the tongue remained the same size, so that the back end was forced further down into the throat and pushed the larynx down with it. It was Negus who first suggested this, but it was still being endorsed in 1974: 'As jaws were severely shrunken and retruded in adaptation to cranial balance, neck structures had to give way to make room for the backwardly shoved oral contents'.[15] One extension of this hypothesis suggests that the backward-shoving may also account for our highly arched palates: the roof of the mouth may have buckled upwards in order to become shorter and match the decreasing length of the lower jaw. But Roger Crinion, who is researching the matter, has found a wide variety of mammals with arched palates, from walruses to various fruit-eating bats, and notes that most of them are suction-feeders.[16]

Nothing forced it

The weak point in the shoving-back theory (and, indeed, in all the others) lies in the statement or indication that '. . . the neck structures *had to* give way' [my italics].

There was no such necessity. When the jawbone grew smaller and the teeth grew smaller, so also the muscles that moved the jaw grew smaller and the skin that covered it grew smaller. Why didn't the tongue grow smaller to fit into the smaller space available for it?

There are other mammals with flattened faces. The ones that spring most readily to mind are those that seem to have been artificially bred to suit our own anthropomorphic conceptions of beauty: the Persian cat and the Pekingese dog. Their oral contents have been shoved backwards with a vengeance; it has led to some respiratory problems, but their neck structures have not had to give way. Why did ours have to?

Sir Arthur Keith, Negus's mentor who had launched him into this specialised area of research, pointed out the snag. Negus wrote: 'Sir Arthur Keith has pointed out to me that if Nature had any necessity for maintaining close relation between nose and larynx, it would find means to do so'.[17] And he himself concurred: 'No doubt the gap between epiglottis and soft palate, as seen in Man, could be obviated by lengthening of the palate and epiglottis, if these were necessary'.

In other words, he conceded that the larynx was not *forcibly* pushed backwards. Whatever the alleged forcing agent—gravity, flexion, face flattening, or any other—it *did not have to happen*. The connection between the larynx and the nasal passages would have been retained if that had been 'necessary'. Also, one might argue, if that had been desirable. And everything that Negus had written about the disadvantages and evils attendant on the changes showed that it *would* have been desirable. He was too scrupulous a scholar not to note the difficulty, and unable to find a way round it. Like Montagna on skin and Pond on fat, he greatly extended our knowledge without finding an answer to the question with which he started out.

If the 'forced-down' theory will not hold, we are back with the necessity of trying to think of some situation in which the descent of the larynx might have afforded some significant and immediate advantage.

The palate

Something else in the back of our throats is untypical of the general run of mammals.

Most of the pictures in the anatomy books dealing with the larynx show lateral cross-sections of the head. But there is another way of thinking about the larynx, and that is by looking at pictures of the palate as seen from underneath. F. Wood Jones examined a number of species from this perspective and wrote a paper about his findings.[18]

There are two sections to the mammalian palate. Nearest the front teeth is the hard palate, roofed with bone. Further back it is continued in the shape of the soft palate, a sheet of tissue which extends backwards in most mammals all the way to the back of the mouth, and is pierced only by the small circular hole through which the larynx passes upwards in the two-tube system. The hole is encircled by a sphincter—a ring of muscle which can contract, close around the root of the larynx and hold it firmly in place as long as the animal is alive.

When the animal is dead the muscle relaxes and the larynx slides out downwards. That is why some anatomists who had repeatedly dissected pigs, for example, refused to believe that the larynx in this species is intranarial. I even had one letter from a butcher contesting this point. Wood Jones pointed out that the argument had been settled by putting a hand into the mouth of a live pig. It is an eminently repeatable experiment.

In live mammals the sphincter relaxes for purposes of

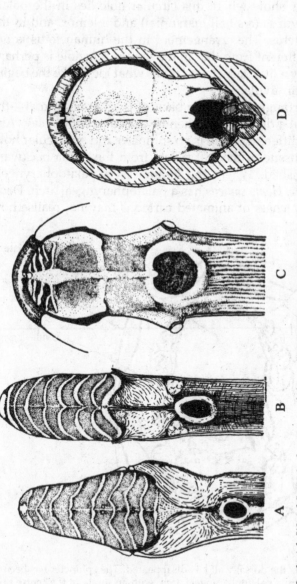

Frederick Wood Jones's drawings of the palate and dorsal wall of (A) dasyurus, (B) lemur, (C) a human embryo of 77 mm. R.V. length, (D) a human embryo of 100 mm. R.V. length. All show the palatopharyngeal sphincter through which the larynx passes up into the intranarial position.

panting or making a noise, or swallowing. Wood Jones's drawings show where this circular hole lies in the palate of a dasyurus (a small marsupial) and a lemur, and in the human fetus. The arrangement in the human fetus is not much different from that in the lemur. The hole is perhaps relatively a little larger, and it has what looks like the beginning of an uvula.

The arrangement in the back of the human throat—the view that a doctor gets when he asks a patient to say 'Ah', is quite different. There is no sphincter and no circular hole. All the tissue of the soft palate from the uvula backward has vanished. Where the lemur has a palatopharyngeal sphincter, *Homo sapiens* has a palatopharyngeal arch. Dedicated watchers of animated cartoons may not realise how

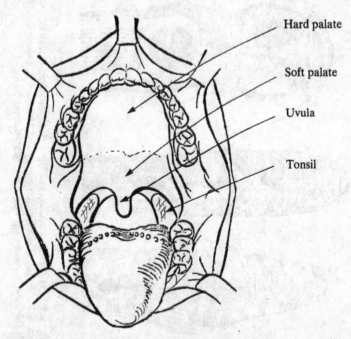

Hard palate

Soft palate

Uvula

Tonsil

In humans the dorsal wall has disappeared. The sphincter has been replaced by a palatopharyngeal arch, with an uvula in the centre of it. *Peter Rhys Evans.*

rare that is. Close-ups of Goofy or Mickey Mouse registering terror always depict a palatopharyngeal arch with a quivering uvula dangling in the middle of it—but that is pure anthropomorphism.

You might imagine from all the talk of the mouth's contents being shoved backwards that this change must have taken place at the same time as the descent of the larynx—that the larynx took a plunge downwards, and carried half of the soft palate down with it, or tore it loose leaving nothing but a gaping hole. That would be a mistake. It preceded the descent of the larynx.

The change from the sphincter to the arch had already happened in the time of the last common ancestor. We can be confident of this because it is a feature we share with other higher primates. Apes have a gaping hole just as we do. They have a dangling uvula just as we do. So does the rhesus monkey.[19]

This provides a really dramatic confirmation of Sir Arthur Keith's contention: that if it was important to keep the larynx in contact with its palate—other things being equal—natural selection would find a way of allowing it to do so, come hell or high water. The ape's larynx has no neat little hole and no sphincter muscle to tighten around the larynx and hold it in place. Nevertheless, its larynx *did not descend*. As Negus described it: 'Chimpanzees, Orangs, and Gorillas have the soft palate only just in contact with the epiglottis, but even in them there is some difficulty in separation of the two, the absence of a hyoid-glosso-epiglottis muscle being foremost among reasons'.[20] Some other factor must have been at work in the case of our own ancestors.

If there was an aquatic interlude, the ape that waded into the water, and later swam and dived in it, already had a gaping hole at the back of its throat. The sole advantage which mouth-breathing bestows on humans is the ability to gulp in large quantities of air very quickly, an ability of particular advantage to a swimmer or a diver re-surfacing after a long breath-held dive. To derive this advantage, an aquatic ape would have needed more than a gaping hole at the back of its throat. It would have needed its larynx to descend to a point low enough to derive the benefit of that gap.

That is admittedly only a speculation. But on this issue even speculations are in very short supply. Some light might be thrown on the matter by inviting a few Olympic water sports champions to complete ten lengths or a couple of dives with mouth taped shut, and seeing how much difference it would make to their performance in the water.

The idea that mouth-breathing is an aquatic adaptation is supported by the fact that terrestrial birds are nose-breathers whereas diving birds are mouth-breathers. The walrus, one of the very few mammals with a descended larynx, is also a mouth-breather.[21]

The power of speech was not bestowed on us by the descent of the larynx. We owe that to another adaptive change in human physiology—one that is even more strongly indicative of a probable aquatic or semi-aquatic phase in our evolutionary history.

12

Why Apes Can't Talk

Syntax overrides carbon dioxide: we suppress the delicately-tuned feed-back loop that controls our breathing to regulate oxygen intake, and instead we time our exhalations to the length of the phrase or sentence we intend to utter.

Steven Pinker[1]

The ability to override carbon dioxide—that is the real pre-requisite of speech. The descent of the larynx was not essential, although one side-effect was an increase in the variety of different sounds we could produce. But as Philip Lieber-mann has stressed: 'Apes do have the anatomy that would suffice to make many of the sounds that occur in speech. The supralaryngeal tract of apes would allow them to produce enough phonetic contrasts to establish a limited speech code'.[2]

So their problem does not lie in their tongues or their lips or their throats. The greatest obstacle does not lie in their brains, either. They are intelligent enough to express them-selves in sign language. They can show their understanding of words and—up to a point—of syntax, by correctly carry-ing out instructions that are given to them, on the lines of: 'Get the yellow ball from the table and put it in the big bucket'. They can answer questions put to them by pointing or pressing buttons. But if you ask them to speak, they cannot get to first base: they cannot say 'Ah' when you ask them to.

Negus expressed this by saying that '. . . primates are highly resistant to the acquisition of speech'. But 'resistance' is hardly the word. They may be ready and willing, but they are not able.

To us, this is such an alien state of affairs that perhaps the best way to understand is to compare it with the kind of instructions that we ourselves would be unable to carry out. The doctor may tell you to breathe in, or cough, but he does not say things like: 'Now, I'd like you to slow down the heartbeat', or 'Kindly sweat into this bottle for me'. Similarly, a man cannot produce an erection to order. Although he may have done it thousands of times before, though he may be anxious to co-operate and be willing it to happen, it is not his to command. An ape is in the same position when asked to utter.

Almost no statement about living animals is true without qualification. K. J. and Caroline Hayes trained their chimpanzee Vicky to be able to utter four words—'papa', 'mama', 'cup' and 'up'. But it took them six years of hard work. (In six years it would be possible to train a man to slow down his heartbeat.) But the people who have been closest to apes all agree that apes' vocal utterances are involuntary expressions of states of mind, not acts of will. Jane Goodall believed that chimpanzees lack the ability to make sounds that are not associated with other displays of their emotional state.[3] R. A. Gardner concurred: 'It is the obligatory attachment of vocal behaviour to emotional state that makes it so difficult, perhaps impossible, for chimpanzees to speak English words'.[4] Philip Liebermann sums it up: '. . . their speech production deficits appear to derive from deficits in motor control as well as anatomy'.[5]

The muscles which the ape cannot consciously control are not in the mouth or the throat, but much lower down. They are the ones that cause the rib-cage to expand and contract, or the diaphragm to move up and down, so that air will flow in and out of the mouth.

Generally speaking, this is not a matter over which mammals need to have conscious control, any more than they need conscious control over the secretion of their gastric juices, or the peristalsis in their intestines. These things are automatically taken care of. If the carbon dioxide level in their blood rises, the muscles controlling the lungs will ensure the inhalation of more oxygen. On the other hand, if the animal falls into the water, or has cold water splashed into its face, breathing will be suspended.

To a large extent our own breathing is under the same unconscious control. We do not have to stay awake all night to ensure that our breathing does not grind to a halt. Sneezing is not a thing we decide to do, and sometimes sobbing or even laughing can get out of hand. The universal response to the splash of water is so independent of the conscious mind that it has been observed to take place in a patient in deep coma.[6]

But, unlike the primates, we can override carbon dioxide by taking a deep breath when the CO_2 level does not require it. We can decide to hold it for a count of ten; we can, when instructed, empty the contents of our lungs into a breathalyser. A singer about to deliver an aria, or an orator about to deliver a peroration, can judge how much breath will be needed to get to the end of it.

If that capacity had not been there *first*, the evolution of the elaborate expertise of our lips and tongues would have been pointless, like a car without petrol; areas of our brains would never have become specialised for the production and the interpretation of words. It must have begun with the lungs. And the only species that share with us the ability to 'override carbon dioxide' are diving species—aquatic birds and aquatic reptiles and aquatic mammals.

A seal or a porpoise can be trained to dive into water to retrieve objects. It can be trained to recognise words referring to different objects—a ring, a square, or a triangle—and retrieve the one requested. In a deep tank, if the objects are placed at different depths and the animal's breath is being measured, it can be shown that the amount of breath it takes before the dive varies according to what the depth of the dive is *going to be*. That is one acid test of breath control: reactions can be governed according to what the animal intends to do, or expects to happen.

The diving reflex

In 1870 the French physiologist Paul Bert made a discovery about ducks when he held one in his hands and submerged it in the water. Anyone who has held a bird is liable to be impressed by how quickly its heart beats, especially if it is frightened or constrained. Bert was surprised to note that the duck's heart slowed down dramatically when it was pushed below the surface and only returned to normal speed when it was brought up again.[7]

Six years later another Frenchman, C. Richet, established that presumably because of the slower heart rate and the reduced need of oxygen, ducks with their heads under water can live without breathing three times as long as ducks with their heads in air.[8] In the 1930s a similar response was found in diving mammals—the muskrat and the beaver.

This slowing down of the heartbeat (bradycardia) is one of the components of what has come to be known as the diving reflex. The other main component is vasoconstriction—the shutting down of the blood supply to the skin and the less vulnerable organs of the body so that the limited oxygen available can be devoted to the most oxygen-dependent organs. Some body tissues can be deprived of oxygen for as long as an hour without suffering permanent damage, while others—particularly the brain—must be assured of a constant supply. Further researches established that the diving response did not need total immersion to activate it: it could be triggered off by simply immersing the head or face of an animal in cold water, or even by applying cold water to the face without actual immersion of the nostrils. And scientists came to realise that these physiological reactions for preserving oxygen were not confined to diving species: they are present in all mammals. That makes sense in Darwinian terms. There is a vast amount of water on the earth's surface, and throughout much of the evolutionary history of every species there must have been the occasional danger of falling into a lake or a river, or the sea, or being overwhelmed by a flood.

In fact some such reaction is older than the mammals, even older than air-breathing terrestrials. If you take a fish or a shark out of water it will take broadly similar precautions to preserve its life until it is returned to its native element—a kind of diving reflex in reverse.

Thus, the fact that the diving reflex is manifested in humans is not in itself either surprising or a proof that our ancestors went through a semi-aquatic phase after the ape/human split. It is an instance of a very general asphyxial defence mechanism common to all vertebrates from fish to man. At first, however, some people were surprised to find it in man, and science journalists tended to refer to it as 'vestiges of the diving reflex'.

They are more than vestiges. When R. W. Elsner investigated the matter in 1970 he reported that face immersion alone—usually for one minute—produced as much response as total body immersion. 'Sometimes the effects were dramatic. In one experiment a normal healthy young man held his breath for 30 seconds after respiration. His heart rate fell precipitously until the longest interval between beats was equivalent to a heart rate of only 13 beats per minute.' The 'longest interval' was perhaps not a very significant measurement, but Elsner went on to say that '. . . three other normal subjects have responded with similar and even lower rates. Limb blood flow was almost completely stopped in some subjects simply by face immersion'.[9]

Most results were more moderate, but as a general rule the bradycardia effect slowed the heart rate to about two-thirds of the normal rate in air, and the vasoconstriction reduced the blood flow in the calves to about one-third. These discoveries were significant enough to lead to some medical applications. Patients suffering from a condition called PAT (paroxysmal atrial tachycardia) had formerly been treated by massage and drug treatment requiring hospitalisation, but it was found—in the University of Dallas—that dunking their heads in ice-cold water reduced the heartbeat from 160 a minute down to a normal 70 or 80 within 15 to 35 seconds.[10]

Another phenomenon which was described as a 'vestige of the diving reflex' was the discovery that human babies could survive longer without oxygen than human adults. Could this be—like their more generous allowance of adipose tissue—a relic of a period of re-adaptation to aquatic life? It was a tempting thought, but it cannot be substantiated. In fact, all young mammals are more resistant to asphyxia than adults of their own species. This was discovered in the case of kittens by the scientist Robert Boyle (famous for Boyle's Law) as long ago as 1670. It probably has nothing to do with water, but a lot to do with the birth crisis. There is always a possible hiatus between the time when the young animals are still receiving oxygen through the umbilical cord and the time when they successfully switch to breathing through their own lungs.

To sum up, the diving reflex is a feature common to all mammals. There was some initial surprise at finding it present and fully functioning in *Homo sapiens*. We now know enough to realise that it would have been far more surprising if it had been absent.

It is more highly developed in aquatic mammals than terrestrial ones, which suggested that comparisons between man and other species might throw some light on our own evolutionary history. The matter turned out to be far more complicated than it looked.

Comparisons

The first attempt to compare diving responses in seals and humans were made by subjecting them to immersion in water for a measured length of time. The results suggested that the reflex in humans was much weaker than that of the seal and other aquatic mammals.

However, the early results had to be reconsidered when it was realised that the first experiments were not comparing like with like. In the case of the human subjects the researchers did not grab hold of any human passing through the laboratory by the scruff of the neck and plunge his head into a tank of water and hold it there. They asked for volunteers. But the animals used by a number of the experimenters were not volunteers. A common procedure in the 1960s was to strap a seal or a dolphin to a board and lower it into water so that the length of the dive could be controlled and timed.

Then R. W. Elsner introduced the practice of training aquatic mammals to dive voluntarily by rewarding them with fish when they dived down into a tank to reach a target.[11]

The fact that the dives were performed voluntarily made a dramatic difference. A sealion's heart rate fell to 40 beats a minute. In the same animal when forcibly submerged— and having no idea when or whether it would be able to take its next breath—the response had been many times as great, with the heart plunging immediately down to eight beats a minute.

P. J. Butler and D. R. Jones commented: 'The diving response, developed during forced submersion of diving animals, seems to be a maximum cardiovascular adjustment, and might never be displayed during the normal life of these mammals'.[12]

This highlighted one of the many factors which makes research in this field so difficult: psychological factors exert a strong influence. A seal's lungs and cardiovascular system react differently to the same physical stimuli *according to what the animal thinks is going to happen next*.

If the animals' responses to immersion had been overestimated, that of humans had been underestimated. Human subjects in early experiments were generally people with little or no training in breath-held diving. *Homo sapiens*, for the most part, has led a terrestrial existence for millions of years. Sudden immersion is as likely to cause panic in an untrained human as it does in a cat or a rabbit, and our children learn by precept or example to regard water as a dangerous element. The fear is partially a trained response. To the layman the most striking demonstration of this in recent years has been the films and photographs of young babies moving freely and fearlessly under the water, clearly enjoying the experience of immersion.

When the researchers began to use trained subjects they found a dramatic difference in the results. Bradycardia in face immersion in untrained humans reduces the heart rate by an average of 25 per cent. In groups of experienced skin divers the average reduction is 45 per cent. Among ethnic groups where diving is an important part of their daily activities, the response is even higher.

How long a period of training does it take to begin to make a difference to our physiological response to submergence? About a quarter of an hour. Most of the early studies in humans had been based on the results from *single immersions*. It has now been found that face immersions repeated at short intervals—sometimes called 'short-term training', demonstrates a rapid adaptation. U. Hentsch and H.-V. Ulmer discovered that breath-holding time could be prolonged by 160 per cent by performing five breath-holds spaced out at three-minute intervals.[13] Intensive training enables divers to continue holding their breath beyond the breaking point where the rise in carbon dioxide prompts a resumption of breathing, rather as long distance runners train themselves to continue past the 'pain barrier'.

The comparison which would be most valuable would be a comparison between humans and other primates, in order to assess the likelihood that human capacities in this respect became modified at an early stage in Africa, following the ape/human split. Objections have been raised that the diving reflex would not have worked at all in those tropical climes because the temperature would not have been low enough to stimulate the cold receptors on the face

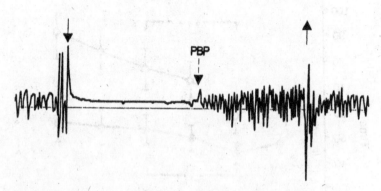

Registration of thoracic movements during an apnoeic episode. Initiation and termination of apnoea are indicated with arrows. PBP indicates the physiological breaking point, the point of occurrence of the first involuntary breathing movements. (Registrations were obtained with a respiration transducer TSD 101 Biopac systems Inc. Goleta, USA). *After E. Schagatay.*

and elicit the reaction response. But Erika Schagatay has researched the question of the cold receptors, located them with greater precision (mainly on the forehead and around the eyes) and ascertained that the determining factor is not the precise temperature, but a difference in temperature. The diving response would be evoked in hot countries as long as the water was colder than the air: 'The present findings suggest that the diving response may very well be operant in a tropical diver as long as his/her face is exposed to air sufficiently warmer than the water between dives'.[14]

Unfortunately, it appears that comparisons between humans and other primates in this field have not been obtained because of the difficulty of inducing primates (and especially apes) to submerge voluntarily. The difficulty in itself may be significant.

There are physiological factors which may contribute to the apes' reluctance. If they submerged they would have

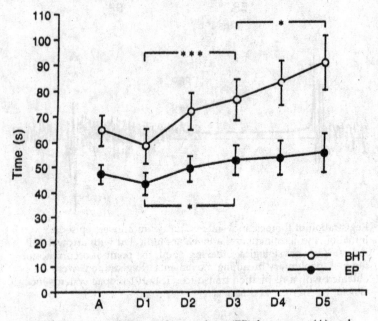

Breath-holding time (BHT) and easy phase (EP) for apnoea (A) and dive 1-5 (D1-D5). Values are means with SE from 23 subjects for which the physiological breaking point could be detected in at least four of the six tests. Significance at the 5% level is indicated. *After E. Schagatay.*

to make greater efforts to return to the surface because their bodies are less buoyant than ours, having a lower proportion of fat. Frank Sandon worked out in 1924 that motionless floating cannot be achieved in fresh water if the specific gravity of the body is more than 0.9875.[15] Human beings, on average, are close enough to that figure to ensure that the air in the lungs makes a crucial difference. Lungs full = floating; lungs empty = sinking. Women are more likely than men, babies more likely than adults, and fat people more likely than thin ones, to fall below that magic figure of specific gravity and float with ease. But the buoyancy factor is greatly dependent on the ability to inflate the lungs and keep them inflated at will—and the apes, as we have seen, do not possess that ability.

As for comparisons with other mammal species, Erika Schagatay—after collating the known facts and conducting her own researches—concluded: 'The diving response of trained human divers is in the range of semi-aquatic mammals like the beaver and the otter, while that of untrained humans is in the range of terrestrial species such as the pig'.[16]

When and why the hominids spoke

It seems a tenable hypothesis that voluntary breath control (the prerequisite of speech) and the descended larynx (which increased the range and variety of sounds it was possible to make) both emerged in the earliest stages of separate hominid evolution and had something to do with water.

This does not mean hominids at an early stage began to speak. But it is possible that sound began to play a somewhat more important part in their communications with one another. Sound is of increased importance to many aquatic mammals. Sounds used for echo-location have evolved independently in species as different as seals and dolphins; this can be more effective than the use of sight in locations where light is limited, such as the ocean bed or in muddy estuaries.

The sense of smell, as well as of sight, can be of diminished usefulness to aquatic mammals. In the extreme case of the cetaceans (whales and dolphins) it has in effect been discarded: no olfactory bulbs have been detected in their brains. (The olfactory bulb in humans is said to be less than half the size of that of the gorilla and the chimpanzee.[17]) An aquatic primate could have had its own reasons for placing increased reliance on the vocal channel of communication.

Social communication in primates, while it includes sounds, is heavily dependent on body language. Apes and monkeys have developed this mode of communication to the point where they can convey their feelings and intentions with great subtlety and effectiveness. Other things being equal, it might be surmised that if an ape at any point needed to improve its communication skills, it would concentrate on refining still further the sign language at which it is so adept.

But this mode of communication between two animals is subject to two limitations. It depends on their being mutually visible; and it depends on their freedom to deploy their bodily postures and movements to convey their messages. In water less of the body is visible (Darwin was struck by the greatly enhanced expressiveness *of the human face*) while postures and movements are largely dictated by the exigencies of keeping afloat. In such circumstances the importance of the vocal channel could be to some extent enhanced, especially once the use of it had been brought under the control of the conscious mind.

Vocal communication, however, is not necessarily speech. The potential for speech may have remained latent for millions of years. Many speculations have been advanced about when and why speech evolved. Suggested reasons for the need of speech have included the need to pass on tool-making skills, and/or a trend towards larger social units demanding more subtle means of communication to regulate them. All of these hypotheses remain unaffected by AAT, which attempts only to throw light on how speech became accessible, not on how it became necessary.

13

Infrequently Asked Questions

Scientists tend not to ask themselves questions unless they already have the glimmerings of an answer.

Sir Peter Medawar

The preceding chapters outline the core arguments for regarding AAT as a tenable hypothesis. They deal with some of the fundamental life processes, such as locomotion, thermoregulation and respiration.

But there are other, apparently less basic, questions which are much easier to shelve, questions like, 'Why are we the only primates with everted lips?' or, 'Why are our babies born covered with grease?' Some academics robustly recommend letting such sleeping dogs lie:

> In order not to unduly restrict the chances of finding possible answers, one should avoid focusing exclusively on traits which seem to be enigmatic. Instead, features should be considered which can be readily understood according to current ideas, or which can be traced through our human ancestry.[1]

It probably makes good academic sense, but anyone dissatisfied with 'current ideas' is more likely to respond to the Sherlock Holmes maxim: 'Tell me everything, omitting no detail, however slight.'

Sex and evolution

'We are the only species that copulates face-to-face,' wrote Jacob Bronowski in 1974.[2] Later he admitted having forgotten the aquatic mammals.

The question of why we mate in this way used to be a hardy perennial, and was often answered by implying that the object was to enable the lovers to gaze into each other's eyes, which made the procedure more significant and helped to cement the pair bond. 'Face-to-face sex is "personalised" sex.'[3]

In my first book (*The Descent of Woman*) I posed the usual AAT question: 'Where else is this behaviour at a premium?' and came up with an increasingly familiar answer: 'In the water'. Whales and dolphins and beavers and many other aquatics use the ventro-ventral approach as humans do, whereas for all anybody knew at the time *Homo sapiens* was the only terrestrial example of this practice.

I argued that both bipedalism and swimming realigned the spine and hind limbs into a straight line, instead of the right-angled arrangement common and convenient in quadrupeds. That redeployment changes the angle of the vagina relative to the hind limbs and makes it more accessible from the front than the back.

This mechanical explanation seemed to be tacitly accepted in place of the alleged social advantages of mutual eye-gazing, and speculation about the problem dried up. It was probably easier to assimilate because it was not actually dependent on acceptance of AAT. As a reviewer in *Nature* commented, the same consequence would have followed from bipedalism regardless of why it evolved.[4]

Meanwhile, on the ventro-ventral front science has moved on. Accounts are coming through of other primates which have been observed mating face-to-face. To date there are only two species involved—the orang-utan and the bonobo (previously known as the pygmy chimpanzee). These discoveries do nothing to encourage a revival

of the pair-bonding theory since neither species is monogamous.

The male orang-utan has difficulty in mating because he
is a large animal and, unlike the gorilla, rarely descends
from the forest canopy. The normal quadrupedal mode
would involve the female standing on all fours gripping a
branch while a male more than twice her size and weight
performed a kind of balancing act perilous to both of them.
So their copulation has been described as '. . . generally ventro-ventral. Usually one or both individuals hung from an
overhead branch, and the female reclined against another
branch'.[5] The orang-utan is clearly a highly specialised
mammal with untypical problems—not a promising model
for *Homo*.

The bonobo is the primate which provides the closest
parallel with our own species. It mates frequently face-to-
face, both in captivity and in the wild, and the vaginal
angle has shifted as markedly as in our own species.

Penis size appears to confirm the conjecture made in *The
Descent of Woman* that increased length of the organ in
humans was not for purposes of display or intimidation or
cementing the pair bond. 'It grew longer for the same
reason as the giraffe's neck—to enable it to reach something
otherwise inaccessible'.[6] Until recently the human male was
believed to have the longest penis of any primate: on average, gorilla 3 cms, chimpanzee 8 cms, man 15 cms.[7] But of
the bonobo—which has also evolved some way towards
ventro-ventral copulation—Frans de Waal wrote
'. . . relative to body size (and probably absolutely as well)
this ape's testicle size and erect penis length surpasses those
of the average human male, until recently thought to be
the champion'.[8]

A second physiological anomaly which has been debated
in connection with AAT is the existence in human females
of the hymen, which is common among aquatic mammals
and a variety of terrestrial ones. But it is not common
among primates. It is present in the lemurs, but not in

monkeys and apes.[9] Presumably there must be some reason why, after such a long hiatus, it was reacquired by the hominids when they became a separate species.

In many aquatic mammals there is a tendency for external organs to be retracted within the body wall and covered up, possibly for purposes of streamlining. In cetaceans and seals this is true, for example, of the ears; in those seals known as 'otaries' the external ears are smaller than in most mammals, and in the other class of seals, the pinnipeds, the process has gone further and the ears are not visible at all. Something analogous seems to have happened to the vagina in human females. Compared with that of apes it is situated more internally and, quite apart from the hymen, it is covered and protected by the fleshy folds known as the labia majora. In the gorilla and the chimpanzee these structures begin to develop *in utero*, but disappear before birth.

The third anomaly in the physiology of human sex is that women's menstruation appears to manifest a lunar rhythm. Chris Knight has discussed this in his book *Blood Relations*, subtitled 'Menstruation and the Origins of Culture'.[10] It deals, among other things, with the phenomenon of synchronised menstruation, and the curious circumstances that the average length of the human menstrual cycle is 29½ days—exactly the same length as a lunar month.

Lunar biorhythms are not unknown in nature. They are found in several species of fish and in some species of frogs and toads. They are commonest in marine creatures: seahorses, for example, only lay their eggs at full moon. And when Sir David Attenborough decided to film horseshoe crabs he had to take into consideration not only the right season of the year to turn up on the beach. He also— like the Christian Church working out the date of Easter— needed to know the phases of the moon. 'Then,' he recorded, 'on three successive nights when the moon is full and the tides are high, hundreds and thousands emerge from the sea.'[11]

When I considered these problems, I could think of no way in which it would have been advantageous for an aquatic ape to synchronise with the tides, unless perhaps on the assumption that the gestation period is a multiple of the menstrual one. In that case, it would be possible to envisage circumstances in which the hazards of giving birth on a beach would be greatly affected by whether the tide was high or low.

Chris Knight's approach is different. He writes: 'If the human menstrual cycle was genuinely linked with the moon it would be rather surprising, for such a correspondence is not normal, either for primates or for mammals in general'.[12] He interprets it in terms of his hypothesis of a sexual strategy: 'If coastal females were beginning to synchronise with each other for their own sexual-political reasons, then any external cues with an appropriate periodicity would automatically have acquired special significance'. That could have stabilised a roughly four-weekly cycle into a more precisely lunar one, and that could have been further reinforced in the savannah period and subsequently by the characteristically lunar periodicity of rituals connected with human hunting—a suggestion documented by accounts of practices and rites in many societies in different parts of the world.

More on the skin

In discussing sweat-cooling, reference was made to the apocrine glands and the eccrine glands, both (in different species) adapted for thermoregulatory purposes from earlier primitive functions.

There is a third type of skin gland which remains even more enigmatic. No one has been able to understand why, in humans, thousands of sebaceous glands—specifically those on the face, head and back of the torso—are so extraordinarily large and become so fiercely active in adolescence. Sebum is an oily fluid whose only known function in mammals is waterproofing the hair and skin. The proliferation of the sebaceous glands in humans is not paralleled in the apes, where the glands are very small and confined to localised sites—surrounding the eyes, the lips and the anus. There is no pre-existing primate trend towards producing more sebum.

William Montagna found the phenomenon puzzling: 'The human body appears to contain useless appendages and even to make mistakes, but the sebaceous glands are too numerous and too active to be described as trivial'.[13]

Another researcher, A. L. Kligman,[14] decided to spit on his hands, investigate in depth and settle once and for all the question of what useful function the sebaceous glands performed. He concluded there is no answer to this question. They serve no purpose. They are an obsolete relic.

Of what?

If the phenomenon is unique to our species it surely calls for an explanation; it could be a relic of a time when our bodies were not only covered with hair but with hair richly endowed with waterproofing protection.

The idea is rendered more probable by the fact that during gestation—when the fetus grows the first coat of hair (lanugo) covering its body, head and face—sebaceous glands all over its body become very active. The sebum exuded at that time remains present after the lanugo has vanished; it is this which constitutes the 'vernix'—the 'cheesy' or 'waxy' substance usually found clinging to the body of a new-born baby.

The probability is that the vernix is unique. We cannot be quite certain because our own babies are born at a relatively immature stage of their development. It is just possible that some such secretion may be exuded by the fetuses of other primates but has never been noticed because it disappears before birth. The trouble with infrequently asked questions is that as long as they remain unasked, the facts which might provide the answers are unlikely to emerge.

Hair tracts

In earlier discussions of the skin use was made of the convenient but imprecise term 'naked'. It should have been
'functionally naked', in the sense that hair over most of the
human body is too short and sparse to fulfil any of the
functions served by the hair of most mammals.

But it is there. And one of the earliest arguments used
in support of AAT by Alister Hardy concerned *the orientation* of the hairs that remain. He pointed out that it is quite
unlike that of other primates, and appears to follow the
direction of the water that flows over a swimming body.
This argument cannot be dismissed out of hand. The orientation of hairs in other mammals is not random. It does as
a rule bear some relation to their life-style, at least in
relation to gravity, and the track followed by any rain that
might fall upon them. Hairs pointing downwards appear
designed to help the rain to run off onto the ground rather
than penetrating to the skin.

The rules affecting this are not clear cut. Some animals—
like camels and moles—are unlikely to have rain to descend
on them and their hairs do not lie down at all. There may
well be other reasons for this. A mole needs to be able
when necessary to move backwards as well as forwards
through the earth without getting its pelt clogged with soil,
and the teddy-bear disposition of the camel's pelt is found
in other (non-desert) species like the woolly monkey. But
the peculiar pelt of the sloth, in which the hairs point from
its belly towards its back—the direct opposite of most
quadrupeds—surely relates to the fact that it lives upside
down.

At the turn of the century the direction of hair tracts was
a fashionable subject of debate among evolutionists. Man,
in this respect as in many others, appears to be unique. 'The
difference,' wrote Walter Kidd in 1900, 'in many respects is
startling'.[15] Two of them are particularly striking. On the
back, the hair of other primates points downwards: 'Briefly

156 *The Aquatic Ape Hypothesis*

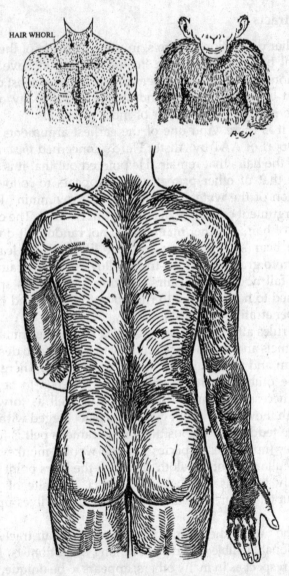

The hairs on an ape's back point downwards. The illustrations, drawn by Walter Kidd, suggested to Alister Hardy that in humans they trace the flow of water over a swimming body.

and roughly it might be said that the trend of the hair stream is from the cephalic to the caudal extremities', and '. . . the more near to the erect position is the normal position of an ape or monkey the more does the hair lie in the horizontal axis on the trunk'. In humans, which retain the erect position for locomotion as well as squatting, the longitudinal disposition should have reached its apogee.

Instead, the hair streams emerging from under the armpits first begin to curve upwards to the neck, then turn sharply downwards towards the spine, converging in a central line which continues downwards towards the coccyx. 'This striking change of slope,' Kidd commented, 'is certainly not inherited from any known member of the Simian family and is almost in direct opposition to that found in all other mammals.'

The second major difference is a parting on the front of the upper thorax, on a straight horizontal line just below the level of the shoulders. The hair above this parting streams upwards and over the shoulders. Below it points downwards. In apes the tracts are downwards all the way. It seems a legitimate speculation by Hardy that in a swimming hominid holding its head above water and performing some approximation to a breaststroke, the water flowing around the body would follow precisely the course indicated by these two anomalous hair tracts found in *Homo sapiens*. And these particular features of hair distribution are not a recent development. They are already present in the lanugo.

The only alternative explanation on offer is Peter Wheeler's, in *The Aquatic Ape: Fact or Fiction?*[16] He observed that the hair tracts lie in direct opposition to the path taken by the natural convection currents of heated air rising up the body—air warmed by contact with the skin, and/or by the hot surface of the sun-baked savannah. He speculated that this counter-orientation would encourage the air to flow inwards towards the skin, where it would aid in the evaporation of sweat, rather than deflecting it over the surface of the hairs.

It is open to question whether directing warmer air (warmer than the surrounding atmosphere) towards the skin would on balance serve to reduce body temperature. Even if this were the case, the convective air flow above the buttocks is not really in direct opposition to the hair streams. And above the waist any explanatory power the hypothesis might possess breaks down because the convective current 'becomes turbulent as the velocity increases over the upper body'.

Ears and noses

'Behaviour does not fossilise' is a palaeontologist's maxim, and in general that applies to swimming and diving behaviour as much as to any other kind. It would be impossible for an anatomist, examining the fossilised remains of a mustelid, to establish whether it led a terrestrial/arboreal life like a polecat or was on its way to becoming an otter.

But in humans, swimming can leave a trace on the skeleton within the life span of a single individual. The evidence is found in bones of the ear; it consists of bony swellings on the external canal near the annulus of the tympanic membrane. These are known as exostoses, and as Peter Rhys Evans commented in 1992, '. . . the relationship between swimming and exostoses is a phenomenon well known to otolaryngologists'.[17] He reported that differences in the incidence of this condition had been noted in Germany as early as 1904, and that in 1935 P. Belgrave had recorded that, while the incidence among his clinic patients overall was only 2.02 per cent, in members of swimming clubs the incidence was as high as 42.8 per cent and virtually all instructors were found to possess exostoses. Other studies have produced similar results. The evidence of exostoses can be affected by age, sex, racial variations and water temperatures, but the consistent finding in all the studies is that no person was found with exostoses who was not a frequent swimmer.

Ear exostoses have been found in Egyptian mummies dating back to 1500 BC, and a 1990 publication by G. P. Rightmire[18] on the evolution of *Homo erectus* discusses variations or abnormalities of the temporal bone and external canal, including one from Lake Ndutu in the Olduvai Gorge consistent with the diagnosis of external auditory canal exostoses. It is reasonable to conclude that at least some of the *erectus* populations who died by the Rift Valley lakes used the water not only for drinking but also for swimming in.

It should not be hard to believe that the practice of swimming survived as late as *H. erectus*; we know from experience that it has survived as late as *Homo sapiens*. People affluent enough to design their surroundings and choose their leisure activities build themselves bathrooms and showers and swimming pools and spend their holidays by the sea. Once there, they display the thalassotropic tendency that fascinated Robert Frost:

> The people along the sand
> All turn and look one way.
> They turn their back on the land
> They look at the sea all day.

Is this observation purely anecdotal? It may not have been quantified in a scientific paper, but it passes a test at least as rigorous—the economic one. Tourist agencies know that their clients will pay more for a Sea View hotel, and the hotelier knows he can mark up accommodation where one window reveals a sector of a level blue horizon. Such a preference strikes us as 'only natural—just human nature'. But why? It is not primate nature. If we were descendants merely of a grounded arboreal ape, our atavistic instincts should be driving up the prices of the Forest

View hotels, and people in their hundreds would set out their deck chairs to gaze at the trees all day.

The human nose

The human nose is a deeply puzzling problem. In most primates the nostrils point forward (Old World primates) or out to the sides (New World primates). In man they point downwards towards the chin, and are covered with a kind of lid supported by a framework of cartilage. The infrequently asked question is, Why?

The lazy answer is the same as the old one on nakedness—that it is merely an optical illusion. What happened, so this argument runs, was that the face receded—notably the jaws, due to a change of diet, social structure, use of tools or other reasons. Meanwhile, the nose did not recede, creating the misleading impression that it had grown longer. No explanation is offered as to why the nose did not recede when everything else did.

The fossil hunters inform us that the nasal spine only arrived at around the time of *Homo erectus*, around 2.5 million years ago. The nasal spine is the short piece of bone between the eyes which—in man but not in other primates—projects forward for a short distance. The remainder of the supporting structure consists of cartilage, which is more flexible. (It has to be flexible for obstetric reasons. The passage of the baby's head through the pelvic ring is a sufficiently tight fit without having to accommodate bony projections. As it is, it has been surmised that the birth process can cause hairline cracks in the cartilage which lead in later life to the condition called a deviated septum— unknown in other animals but fairly common in humans.)

The late arrival of the nasal spine in the fossil record is no evidence that the nose itself had not begun to evolve earlier. Flesh and cartilage do not fossilise. If elephants were extinct, we would never suspect from their fossilised bones that they possessed trunks. An elephant's skull presents a profile at least as flat-faced as our own. The likelihood is that the nasal spine in humans, necessarily curtailed as it is, evolved as partial support for a structure which was already in existence.

There has been remarkably little speculation about the purpose of the human nose. Negus considered it rather maladaptive; it means that the current of air has to describe something of a U-turn before reaching the lungs, but it does serve a useful purpose for humans who dive head first into water, or swim beneath the surface. It prevents the water from being forced up into the nostrils—as would most certainly happen to a gorilla if it attempted to do either of these things. In humans, the bridge of the nose acts like a prow, and the water is deflected to either side of it.

Finding a possible parallel in other species is very difficult. The obvious candidate seems to be the proboscis monkey. However, the proboscis does not dive into water head first. One troop of 18 or 20 observed by James Kern '...fell in an ungainly spreadeagle position and belly flopped onto the water with a resounding smack.'[19] This monkey's popular name, as well as its scientific name (*Nasalis*) derives from the extraordinary nasal appendage of the adult male. That is, in the male, obviously epigamic. Like other epigamic adornments—the lyre bird's tail, for example—it has got out of hand, driven by sexual selection to absurd lengths. But the proboscis's nose, like many epigamic adornments, represents an exaggerated form of a characteristic which is present in, and peculiar to, the species as a whole.

In the females and the young of the proboscis monkey the nose, though far less exuberant, is still the most remarkable feature. A clue to its evolutionary development can be detected when the young learn to swim. Their upturned noses project above the water like miniature snorkels. Derek Ellis has speculated that the species-characteristic

When the young proboscis monkey swims, its uplifted nose keeps its nostrils above water. *Amanda Williams*

The proboscis monkey on the right was drawn in 1848 by Frank Marryatt. He wrote that it was '. . . very young, and with nose more or less prominent, and giving its face a more actual resemblance to that of a man's than I had ever before seen'.

nose could have originally been '. . . an infant adaptation subject to allometry (continued disproportionate growth) and modified by female sexual selection'.[20]

The incipient snorkel adaptation has been carried a step further in the tapir (a semi-aquatic relation of the horse), and furthest of all in the elephant. The proboscis in these animals is formed by a fleshy cover over the nose, growing downwards, and combining with an everted upper lip growing upwards. The two combine and merge to form a breathing tube. It is an instance of convergent evolution, since the elephant and the tapir—totally unrelated—have evolved this structure quite independently.

I am now warily approaching a very tentative speculation. The following exchange was published in the Question and Answer feature in *The Guardian*:

QUESTION: *If it has a name, what do we call the narrow
channel running vertically from the base of the nose to the
edge of the centre of the upper lip?*

ANSWER: This embryological 'seam' is called the philtrum, and derives its name from the Greek *philtron*, a
potion used to excite love or passion. How the word
became transferred to an anatomical structure is not
clear, but Gray's *Anatomy* refers to the 'philtrum of
the lip', suggesting at least the possibility of philtra
elsewhere. *Dr Stephen Marriage, Department of Paediat-
rics, Imperial College of Medicine, London, W2.*

No other primate has anything like it. No other primate
mouth has evolved everted lips, with a dip in the centre
of the top one, creating what was once known romantically
as the 'Cupid's bow' shape.

Over twenty years ago, coming out of a BBC studio after
a live audience-participation discussion on *The Descent of
Woman*, I was approached by a young student who told
me he could and did close his nostrils when swimming
underwater, and he pushed up his upper lip to show me
how. I paid little attention. I tried it myself, but it didn't
work.

Years later I received a letter from Virginia, which ran
in part as follows:

Upon reading the subchapter 'Aquatics Close Their
Nostrils . . . ?' I was surprised to find no documentation
of humans doing this, which indeed they do. At least
my daughter, who was 'swimming' around in Lake
Champlain before her first birthday—I kept hold of
her diaper and let her do her thing—was discovered
some years later to firmly close her nostrils with her

upper lip while swimming underwater. When this was discovered, she was quite amazed that everyone didn't do the same.

When this 'freakishness' of hers came up in the conversation some weeks ago, it turns out that our son can and does do the same thing, and my husband found to his amazement that he, too, could do it—not me!

Just thought this might be of some interest—you might even like to pursue it—my two kids can't be the only ones.

Bergliot G. Sleight

Years later still, in 1995, I received a letter from St Paul, Minnesota:

Ever since I could swim I have loved to be underwater for as long as possible. However, if one turns on the back in the water, the air bubbles in the nose escape, and give the unpleasant sensation of a nose filled with water. This also led in those days to the odd sight of girls jumping in the pool with their noses pinched.

I learned, however, to press my upper lip against my nostrils, and this seals perfectly in my case. The reason why I bring up this personal habit is because I always wondered why my upper lip seemed to be perfectly made for this. The two lines descending from the nose plug up the holes, while the recess in the middle allows for the bridge between the holes (see drawing).

As far as I can tell this particular feature of our nose is not copied in chimpanzees or gorillas. That is why, after reading your book, I felt an urge to tell you about this use I found for my anatomy, and I wondered whether there is any explanation for it.

At the bottom of the letter the writer appended three pencil

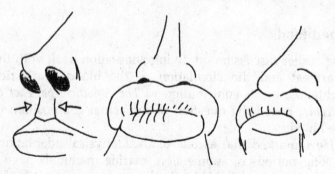

'I learned, however, to press my upper lip against my nostrils, and this seals perfectly in my case . . .' *March 1995, Peter van de Graaf, pers. comm.*

sketches which made it clear what he was describing.

Three unsolicited pieces of testimony at quite long intervals represent too meagre a sample to be significant. But it is slightly reminiscent of the ability to curl the tongue which is found in a minority of people. Presumably it has always been there; presumably, those who could do it never mentioned it because they assumed that everybody could do it. It only became generally known when a scientist became curious and conducted a survey and discovered that the ability is genetically inherited.

In the case of tongue-curling it is hard to decide whether it is an acquired anomaly or a once general condition that has been largely lost. If the ability to close the nostrils proved to be a similar case, it would be easier to guess whether it had been acquired or retained—because every one of us is born with a philtrum. One of the commoner challenges to supporters of AAT is: 'If it was true, why did our ancestors never acquire the ability to close the nostrils?' Just possibly they did, but for land-dwellers it ceased to be adaptive and there was no natural selection pressure to retain it.

The blood

The earlier discussion on diving adaptation dealt with the heartbeat and the circulation of the blood. Karl-Erich Fichtelius, in his contribution to *The Aquatic Ape: Fact or Fiction?*, raised the question of the actual composition of the blood.[21]

He remarked that as one of their physical adaptations to long periods of submersion, marine mammals have a reduced number of red blood cells per unit volume: 'In this respect it is remarkable that chimpanzees have an average of 7.3 million red cells per cubic millimetre of blood, gorillas 6.3 and humans only 5.1 million.'

Also in aquatic mammals there is a higher haemoglobin content per unit volume than is found in land mammals of comparable size: 'The percentages of haemoglobin per cell are about 12.2 for chimpanzees, 13.2 for gorillas, and 18.6 for humans.'

Did the hominids eat fish?

There has always been much speculation by palaeontologists concerning the diet of the early hominids. Most of the discussion hinged on the evidence provided by their teeth. Attention focused on the disappearance of the canines, the shortening of the jaws, the shape and size of the molars, the relative thickness of the enamel and the nature of the wear and tear on the teeth of adult specimens. It was a natural field of research because bones and teeth are all that remain of the australopithecines. Of the two, the teeth survive longer than the bones and in a better state of preservation.

Much of this evidence cast doubt on the original Raymond Dart theory of a major shift from herbivorous to a carnivorous diet, once supposed to coincide with the ape/human split and a move to a hunting life on the savannah. Reference has been made earlier to the seed-eating hypothesis; other suggestions have included scavenging for carrion or collecting fruit. Behind most of the suggestions was a tacit assumption that the sources of the food from which the hominids had to select their diet were all terrestrial in origin.

In the 1980s a new approach to the problem of brain growth was initiated by Robert Martin.[22] He discussed not why it may have become necessary, but when and why it became possible. He pointed out that the brain, compared with most organs of the body, is a particularly greedy consumer of energy resources. Its growth could only have been initiated in conditions where there was an abundant and non-seasonal supply of food, especially for pregnant and nursing females. It was they who had to provide the fuel for brain growth *in utero* and in infancy, when the most rapid cranial expansion takes place.

This looked like the kind of information which AAT could well assimilate. On the savannah, food supplies of all kinds tend to be seasonal and for part of the year distinctly meagre, whereas the coastal waters of the continental shelf in tropical latitudes sustain the richest biomass of any environment in the world, and are not affected by periods of dearth. Primates in coastal habitats readily exploit these resources—eating crabs, digging up turtle eggs, picking limpets off the rocks and sucking out the contents, and beachcombing for dead fish or squids washed up on the shore.[23]

This scenario was reinforced by Michael Crawford in his book *The Driving Force*, where he demonstrated that the rapid expansion of the human brain demanded not only a rich food supply, but one in which there was a correct balance between two different types of fatty acids. Brain tissue in all mammals is unlike other tissues in that it demands a 1:1 balance of Omega 3 (long chain) and Omega 6 (short chain) fatty acids. The Omega 3 type is plentiful in the sea food chain, but scarce in the land food chain, especially in the interiors of continents.[24]

Awareness of the importance to human health of the long chain fatty acids has been reinforced by data on the connection between high seafood consumption (as in Japan) and low incidence of hypertension. Michel Odent's investigations have underlined the importance of these elements in the diet during pregnancy.[25] To protect against heart disease, bread containing fish oil is already on sale in Denmark, and milk with fish oil can be bought in New Zealand.

The difficulty in fitting the new data into the aquatic hypothesis seemed to be one of timing. Firstly, in the AAT scenario the semi-aquatic phase is assumed to have predated Lucy and been a contributory factor to her bipedalism. But it is often pointed out that the big surprise about Lucy was her small brain—no bigger than a chimpanzee's. Why had her brain not begun to expand if her diet had become richer? Secondly, the really rapid period of brain growth began around 2.5 million years ago, and much of the evidence for it is found in fossil hominids living by the shores of the inland Rift Valley lakes, quite a long way from the ocean.

However, the suggestion that Lucy's brain had not begun to expand is too facile. The comparison between her brain size and a chimpanzee's is not comparing like with like. It is comparing the brain size of a hominid 3.5 million years ago with that of a modern chimpanzee. As Robert Martin has pointed out, there is a tendency for gradual growth in relative brain size over time in primates.[26] If Lucy's brain was the size of a modern chimpanzee's, it must have been appreciably larger than those of the apes who were her contemporaries, though its growth rate remained very modest for the ensuing million years.

As for the lake-dwellers during the period of rapid brain expansion, investigation of the link between nutrition and brain size has been broadened to examine the lipid profile of freshwater fish, with specific reference to those of the Rift Valley lakes. A paper by Broadhurst, Cunnane and Crawford concludes that:

The diverse alkaline-freshwater fish species within those lakes provided, either directly or indirectly, a source not only of protein, but also essential polyunsaturated fatty acids (PUFA). In particular, the freshwater fish lipid profile has a ratio of docosahexaenoic acid to arachidonic acid (DHA/AA) that is closer to the ratio in our own brain phospholipids than any other food source known.[27]

It is a new area of research and there is a long way to go. But since there is strong evidence that the hominids' habitat was mainly, if not exclusively, by the waterside (whether salt water or fresh), it is slightly surprising that in all the speculations of what they lived on, aquatic resources are so rarely mentioned.

The baboon marker

In the late 1970s G. J. Todaro published a series of papers which, if accepted, would revolutionise common assumptions about human evolution as radically as they had been upset by the papers of Sarich and Wilson in the 1960s. Sarich and Wilson had suggested that the accepted wisdom was wrong about the *timing* of human emergence—it was far more recent than generally believed. After long debate and resistance their arguments were accepted as valid. Todaro proposed, even more startlingly, that the conventional wisdom was wrong about *the place*. His ideas were greeted with a professional silence that has persisted to this day.

His 1980 paper was entitled 'Evidence using viral gene sequences suggesting an Asian origin of man'[28] and I discussed his conclusions in the last chapter of *The Scars of Evolution*. To summarise it briefly: He presented strong evidence that at one period of prehistory the direct ancestors of *Homo sapiens* were not present on the mainland of Africa.

An endogenous virus, spontaneously arising in baboons

and not harmful to them, was so damaging when it crossed the species barrier that all primates native to Africa carry a retrovirus, which protected their ancestors from the baboon infection when it was at its most virulent. *Homo*, unlike the chimp and the gorilla, does not carry the retrovirus. That indicates that *Homo*'s ancestors were elsewhere when the baboon plague first broke out.

'Elsewhere' means not on the continental mass of Africa at all. The evidence suggests that the virus was airborne. It affected the galagos in the forest canopy as well as the gorillas on the forest floor and the monkeys in the trees and on the savannah. The Congo river was wide enough to prevent interbreeding between the chimpanzees on either side of it and lead to the species differentiation between the common chimpanzee and the bonobo. But it was not wide enough to prevent the spread of the baboon virus which affected primates throughout the whole continent. Microbiology cannot tell us specifically when this event occurred. Todaro himself hypothesised that it happened quite late, and that the human race must be descended from a strain that had emigrated to Asia prior to the baboon plague, and that their descendants moved back again when the danger was over.

Although we cannot date the baboon plague precisely, we can date it relative to other evolutionary events. For instance, it must have happened subsequent to the ape/hominid split. If it had happened in the days of the last common ancestor, then either apes and humans would both have the 'baboon marker' (the retrovirus), or both of them would lack it.

It is also pretty safe to say that it happened previous to the split between *Papio* and *Theropithecus* (that is, between baboons and geladas). Both of these carry the original endogenous virus—not the retrovirus. The chances that it originated in both species independently, after they had separated, is vanishingly small.

So, on Todaro's data we can surmise that the plague

broke out in the window of time between the ape/hominid split and the later *Papio/Theropithecus* split. And at the time when it happened, the protohominids ancestral to modern man were sufficiently separated from the other African primate fauna to have escaped the virus which affected all the others on the continental mainland. So the date of the *Papio/Theropithecus* split is crucial. In 1979 J. E. Cronin and W. E. Meikle submitted a paper entitled 'The Phyletic Position of *Theropithecus*: congruence among molecular, morphological and palaeontological evidence'.[29] They concluded that the separate *Theropithecus* lineage has been in existence for more than four million years, which takes it back to before the exodus of *H. erectus* from Africa and indeed before the emergence of *H. erectus*—back to the time before Lucy, the time of the suggested semi-aquatic interlude.

That time-scale helps to give credence to the speculation advanced by L. P. LaLumiere that a population of hominids could have been marooned on an offshore island.[30]

If that happened, it is likely that any troops of the last common ancestor, cut off in that way, would either have perished or become fairly rapidly water-adapted, taking at least part of their food from the sea. They would have been far enough off shore to escape the infection of the baboon plague and only returned to the mainland after it had lost its virulence.

By far the commonest cause of speciation (the splitting of one lineage into two or more separate species) is a period of geographical isolation. Speciation without separation is so rare that Ernst Mayr doubted whether it really occurred, arguing that '. . . the burden of proof . . . rests on supporters of this alternative mode of speciation'.[31] So there has always been an inherent possibility that the protohominids may have divided from the apes because they were living in a different place. What Todaro's papers have done is to promote the probability of geographical isolation into a virtual certainty.

Among larger mammals, the commonest cause of geographical isolation is a sheet of water. One example is the Congo river which divides the habitat of the bonobo from that of the common chimpanzee. The same process has caused countless island species to evolve—as in Madagascar—along different lines from those of their nearest relatives on the mainland. Again Todaro's discovery, by identifying a feature common to all mainland African primates (but not found in man) reinforces the likelihood that it was a stretch of water which originally divided the protohominids from the apes.

Afar

If there is any validity to the assumption that 'something must have happened' to deflect the protohominids along their eccentric evolutionary path, then Afar merits consideration as the possible site of these events. Of all the places in Africa, Afar was the one where the most violent things were happening—and kept on happening.

Throughout most of the Rift Valley the tectonic movement opening up the rift was a very gradual process—possibly in the order of a mile in half a million years. It is impossible to conceive that this had much effect on the local fauna. It affected them only in the very long term, and then only because it was accompanied by gradual changes of climate and ecology, especially on the eastern side. Similarly, when lakes such as Turkana periodically expanded to double their size, that is unlikely to have been an overnight event that would leave groups of primates marooned and separate from the rest of their species. It may well have been barely perceptible in the life-span of an individual. Any anthropoids in the vicinity could have retreated before the rising waters without realising that they had retreated.

But the Afar Triangle was different. It was one of the locations which Alfred Wegener often cited in support of his theory of continental drift. He believed that Arabia had broken away from the African continent and moved north-east. As for Afar, '. . . if one cuts this triangle out,' he wrote, 'the opposite corner of Arabia fits perfectly into the gap'.[32] He argued that the existing terrain of the triangle is composed entirely of recent lava beds which have welled up to fill the empty space. The process still continues via a chain of seven active volcanoes.

Geologist Harrison H. Schmidt, on his way to the moon in 1972, looked back at the earth and told the flight control-ler in Houston: 'I didn't grow up with the idea of drifting continents and sea floor spreadings, but I tell you, when you look at the way the pieces of the north-eastern portion of the African continent seem to fit together, separated by a narrow gulf, you could almost make a believer of anybody'.[33]

Today the rifting is accepted as a fact. Actually three rift systems converge on Afar—one from the Red Sea, one from the Indian Ocean, and one from Africa. And the consequent incursion of the sea was not a one-off event. Walter Sulli-van, writing in 1974 about the Miocene (between 5 and 20 million years ago) concluded: 'It thus appears that both the Red Sea and the Mediterranean repeatedly dried up during the same period, building up successive layers of salt (from the desiccation) and shale (from the periods of flooding)'.[34]

On the occasions when the Atlantic broke through the Straits of Gibraltar, the waters would crash through into the Mediterranean depression with the force of a hundred Niagaras. Such catastrophic events have continued into recent times. Within the last couple of years it has been discovered by Walter Pitman and Bill Ryan[35] that the Medi-terranean broke through the Bosphorus into the Black Sea less than eight thousand years ago, and that it happened, as Pitman described it, 'in one whoosh'.

The floodings of Afar are liable to have been in the same way catastrophic rather than gradual. As Sullivan points out, 'Like Gibraltar, the Strait of Bab el Mandeb is a narrow potential floodgate'. Some experts on the area, like Haroun Tazieff, believe that the present absence of water in the Afar Triangle is only a temporary phase and that in the long run it will disappear again beneath the sea.

In that area, if anywhere, populations of forest-dwelling apes could have found their habitat transformed very suddenly, could have escaped by the skin of their teeth, and survived only by radical changes in their life-style which would over time produce equally radical changes in their physiology. Nor would it be hard to believe that they ultimately found their way back to the mainland of Africa and migrated south along the chain of Rift Valley lakes and rivers. In 1969 a three-day symposium on the Red Sea was organised by the Royal Society in London, and the contributors found no difficulty with the idea that during periods of desiccation man's primitive ancestors could, on occasion, have walked across the Red Sea when it was a plain of scorching salt flats. The same must have been true of the Red Sea's extension into the Afar depression, part of which is still a plain of scorching salt flats at the present day.

However, the last word in these matters lies with the fossil hunters. If the fossil evidence clearly showed that the human story began in South Africa as was once believed, any arguments about events in the Afar Triangle would be irrelevant.

But ever since the discovery of Lucy, Hadar has had a strong claim to be regarded as the possible cradle of mankind. That claim was powerfully reinforced in 1994 with the report in *Nature*[36] of the discovery by Tim White and his team of *A. ramidus*. 'Most ancient human came from Afar,' proclaimed the headlines.[37]

Fossil remains of 16 or 17 individuals were discovered at Aramis, in the Afar depression. The bones date from around 4.4 million years ago, 800,000 years before Lucy, and are the nearest that anyone has got so far to the last common ancestor. They were the final proof, as Tim White pointed out at the time, that it '... was not the savannah that forced us along the evolutionary road'.

In the same issue of *Nature*, Giday WoldeGabriel and his colleagues reported on the habitat in which *ramidus* had lived.[38] They concluded from the abundance of fossil wood and arboreal seeds that the Aramis hominids lived and died in a woodland setting. *The New York Times* summarised their conclusions: 'Fossils of the oldest human ancestors have been discovered in Ethiopia, where the ape-like creatures lived 4.4 million years ago on a forested flood plain'.[39]

Some features of the bones suggest that *ramidus* was at least incipiently bipedal. If that proves to be the case, it would mean that the original environmental crisis precipitating the move to bipedalism occurred earlier still, but probably not too far away from that forested flood plain.

Tim White's discovery is compatible with AAT both in terms of location and of ecology. In respect of the savannah theory, *ramidus* was the last nail in its coffin.

14

Beyond Belief

The questions posed by the Aquatic Ape Theory are important and valid. The answers it offers are speculative, but no more so than those of any other available model.

It is now generally agreed that the last common ancestor of apes and men lived in Africa in a landscape which was a mosaic, a mixture of trees and grassland.

One sub-group of these animals—for some reason—began to change. First, they stood up on their hind legs and began to walk bipedally; at some point the hair on their bodies changed direction and ultimately they became functionally naked; the larynx descended and was relocated below the tongue; they became fatter; forgot how to pant; lost their apocrine glands and much of their sense of smell; their sebaceous glands proliferated; their nostrils pointed in a new direction; finally they evolved larger brains, gave birth to more immature babies, and learned to speak.

The mosaic theory implies that these changes were set in motion because the hominid's ancestors chose to live in, or occasionally had to cross, the open spaces between the forested areas. Open spaces have never caused any of these changes in any other mammal.

That explanation is not good enough. No one is very happy about this state of affairs, or wishes it to persist any longer than necessary. In July 1996 an article in the American magazine *Discover* predicted: 'For all the effort it has taken to bring down the savannah hypothesis, it will take much more to build up something else in its place'.

This book is offered as a contribution to that building process.

References

CHAPTER 1

1 Tobias, P. V., in the Daryll Forde Memorial Lecture, University College London (Department of Anthropology) 4 November, 1995.

2 The Darwin Seminars, London School of Economics, 14 March, 1996.

3 Dart, Raymond (1953). The Predatory Transition from Man to Ape. Quoted in Richard Leakey and Roger Lewin (1977), *Origins*, p. 208, London: Macdonald & Jane's.

4 Ardrey, Robert (1967). *The Territorial Imperative*, London: Collins.

5 Hall, K. R. L. and de Vore, Irven (1965). Baboon Ecology. In I. de Vore (ed.), *Primate Behaviour*, New York: Holt, Rinehart & Winston.

6 Pfeiffer, John (1970). *The Emergence of Man*, London: Thomas Nelson & Sons.

7 Leakey, Richard and Lewin, Roger (1977). *Op. cit.*, p. 39.

8 Bronowski, J. (1974). *The Ascent of Man*, London: BBC, p. 26.

9 Wheeler, Peter (1991). The thermoregulatory advantages of hominid bipedalism in open equatorial environments, *J. Hum. Evol.*, **21**, 107–15.

10 Wheeler, Peter (1991). The influence of bipedalism on the energy and water budgets of early hominids, *J. Hum. Evol.*, **21**, 117–36.

11 Susman, Randall L. (1987). Pygmy chimpanzees and

common chimpanzees: models for the behavioural ecology of the earliest hominids. In W. G. Kingsley (ed.), *Evolution of Human Behaviour: Primate models*, p. 73, Albany: State University of New York Press.

12 Vrba, E. S. (1974). Chronological and ecological implications of the fossil Bovidae at the Sterkfontein Australopithecine site, *Nature*, **250**, 19–23.

13 Carroll, Robert L. (1988). *Vertebrate Paleontology and Evolution*, p. 475, New York: W. H. Freeman.

14 Leakey, Richard and Lewin, Roger (1992). *Origins Reconsidered*, p. 85, London: Little, Brown & Co.

15 Kingston, John D., Marino, Bruno D. and Hill, Andrew (1994). Isotopic evidence for neogene hominid paleoenvironments, *Science*, **264**, 955–9.

16 *New York Times*, 17 May, 1994, p. C1.

17 Wood, Bernard (1996). Book review in *Nature*, **379**, 687.

18 Hunt, K. D. (1994). The evolution of human bipedality: ecology and functional morphology, *J. Hum. Evol.*, **26** (3), 191.

19 Tobias, Phillip (1995). *Op. cit.*

20 Coppens, Yves. East Side Story, *Scientific American*, May 1994.

21 Dunbar, Robin (1995). *The Trouble with Science*, pp. 23–4, London: Faber & Faber.

22 Westenhofer, Max (1942). *Der Eigenweg des Menschen*, Berlin: Mannstaedt & Co.

23 Hardy, Alister (1960). Was man more aquatic in the past?, *New Scientist*, 642–5.

24 Hynes, H. B. N. (1960). *New Scientist*, 889.

25 Groves, Colin P. (1993). Book review in *Human Biology*, **65**, 1038–40.

26 Dennett, Daniel C. (1995). *Darwin's Dangerous Idea: evolution and the meanings of life*, p. 244, New York: Simon & Schuster.

CHAPTER 2

1   Dart, Raymond (1925). *Australopithecus africanus*: the ManApe of South Africa, *Nature*, **115**, 196.

2   Johanson, Donald C. and Edey, Maitland A. (1981). *Lucy: The Beginnings of Humankind*, London: Granada Publishing.

3   Johanson, D. C. and Taieb, M. (1976). Plio-Pleistocene Hominid Discoveries in Hadar, Ethiopia, *Nature*, **260**, 293–7.

4   Klein, Richard G. (1989). *The Human Career: Human Biological and Cultural Origins*, p. 137, University of Chicago Press.

5   Foley, Robert (1987). *Another Unique Species: Patterns in human evolutionary biology*, p. 195, Harlow: Longman.

6   Leakey, Meave (1995). The dawn of humans, *National Geographic*, **188**, 42.

7   Johanson, D. C., Taieb, Maurice and Coppens, Yves (1988). Pliocene hominids from the Hadar Formation, Ethiopia (1973–1977): Stratigraphic, chronological and palaeoenvironmental context, with notes on hominid morphology and systematics, *Am. J. Phys. Anthrop*, **57**, 373–402.

8   Mohr, Paul (1978). Afar, *Ann. Rev. Earth. Planet. Sci.*, **6**, 145–72.

9   Hay, R. L. and Leakey, M. D. (1982). The fossil footprints of Laetoli, *Scientific American*, **246** (2), 38–45.

10 Andrews, P. J. (1989). Palaeoecology of Laetoli, *J. Hum. Evol.* **18**, 173–81.

11 Willock, Colin (1974). *Africa's Rift Valley*, p. 43, the 'World's Wild Places' series, London: Time-Life International.

12 Johanson, Donald C. and Edey, Maitland A. (1981). *Op. cit.*, p. 151.

13 Martin, Richard Mark (1977). *Mammals of the Sea*, London: B. T. Batsford.

14 Zihlman, Adrienne, Pygmy chimps, people and the pundits, *New Scientist*, 15 November, 1984, 39–40.

15 Jablonski, Nina G. and Chaplin, George (1993). Origin of terrestrial bipedalism in the ancestors of the Hominidae, *J. Hum. Evol.*, **24** (4), 273.

CHAPTER 3

1 Johanson, D. C. and Edey, M. A. (1981). *Lucy: The Beginnings of Humankind*, p. 306, London: Granada Publishing.
2 Morris, Desmond (1967). *The Naked Ape*, p. 21, London: Jonathan Cape.
3 Leakey, L., Tobias, P. and Napier, J. (1964). A new species of the Genus Homo from Olduvai Gorge, *Nature*, **202**, 7–9.
4 Tuttle, R. H. (1975). Parallelism, brachiation and hominid phylogeny. In W. P. Luckett and F. S. Szalay (eds.), *The Phylogeny of the Primates*, New York: Plenum.
5 Aiello, Leslie (1981). Locomotion in the Miocene hominidea. In C. B. Stringer (ed.), *Aspects of Human Evolution*, vol. XXI, pp. 63–79, London: Taylor & Francis.
6 McHenry, Henry M. (1982). The pattern of human evolution: Studies on bipedalism, mastication and encephalisation, *Ann. Rev. Anthrop.*, **11**, 151–73.
7 Kimura, Tasuku (1985). Bipedal and quadrupedal walking of primates: comparative dynamics. In Shiro Kondo (ed.), *Primate Morphology, Locomotor Analyses and Human Bipedalism*, p. 95, University of Tokyo Press.
8 Napier, J. R. and Napier, P. H. (1985). *The Natural History of the Primates*, p. 163, London: British Museum.
9 Tuttle, R. H. and Watts, D. P. (1985). The positional behaviour and adaptive complexes of Pan gorilla. In Shiro Kondo (ed.), *Primate Morphology, Locomotor Analyses and Human Bipedalism*, pp. 261–88. University of Tokyo Press.
10 Wood Jones, F. (1948). *Hallmarks of Mankind*, London: Edward Arnold.
11 Kelley, Jay (1992). The evolution of apes. In S. Jones

(ed.), *The Cambridge Encyclopedia of Human Evolution*, Cambridge University Press.

12 McHenry, Henry M. (1982). *Op. cit.*, p. 155.

13 Jungers, William L. (1988). Relative joint size and hominid locomotor adaptations with implications for the evolution of hominid bipedalism, *J. Hum. Evol.*, **17**, 17, 247–65.

CHAPTER 4

1 McHenry, Henry M. (1982). The pattern of human evolution: Studies on bipedalism, mastication and encephalisation, *Ann. Rev. Anthrop.* **11**, 156.

2 Wood, Bernard (1993). Book review in *Nature*, **363**, 587.

3 Hunter, John. Quoted by Wood Jones, F. (1948) in *Hallmarks of Mankind*, London: Edward Arnold.

4 Pfeiffer, John (1970). *The Emergence of Man*, p. 45, London: Thomas Nelson & Son.

5 Quoted in Johanson, D. C. and Edey, M. A. (1981). *Lucy: The Beginnings of Humankind*, p. 309, London: Granada Publishing.

6 Lovejoy, C. O. The evolution of human walking, *Scientific American*, November 1988, p. 118.

7 Gould, S. J. The upright ape, *New Scientist*, 6 September, 1979, p. 739.

8 Napier, John (1967). The antiquity of human walking, *Scientific American*, **216** (4) pp. 56–66.

9 Wood Jones, F. (1948). *Op. cit.*

10 Morgan, Elaine (1990). *The Scars of Evolution*, chap. 3, London: Souvenir Press.

11 Carrier, David R. (1984). The energetic paradox of human running and hominid evolution, *Current Anthropology*, **25** (4), 483.

12 Lewin, Roger (1987). *Bones of Contention: Controversies in the Search for Human Origins*, p. 40, New York: Simon & Schuster.

13 Taylor, C. Richard and Rowntree, V. J. (1973). Running

on two or four legs: Which consumes more energy?, *Science*, **179**, 186–7.

14   Rodman, Peter S., and McHenry, Henry M. (1980). Bioenergetics and the origin of hominid bipedalism, *Am. J. Phys. Anthrop.* **52**, 103–6.

15   Fedak, M. A., Pinshow, B. and Schmidt-Nielsen, K. (1974). Energy costs of bipedal running, *Am. J. Physiol.*, **B227**, 1033–44.

16   McHenry, Henry M. (1982). *Op. cit.*, p. 163.

17   Aiello, Leslie and Dean, Christopher (1990). *An Introduction to Human Evolutionary Anatomy*, p. 246, London: Academic Press.

CHAPTER 5

1   Anderson, Connie M. (1992). Book review in *Am. J. Phys. Anthrop.*, **88**, 564–6.

2   *Meerkats United*. Video. BBC Wildlife Specials.

3   Lovejoy, C. O. (1981). The origin of Man, *Science*, **211**, 341–50.

4   Napier, John (1967). The antiquity of human walking, *Scientific American*, **216** (4), 56–66.

5   Newman, R. W. (1970). Why man is such a thirsty and sweaty naked animal, *Human Biology*, **42**, 12–27.

6   Bromage, T. (1985). Re-evaluation of the age of death of immature human fossils, *Nature*, **317**, 525–7.

7   Bauer, Harold R. (1977). Chimpanzee bipedal locomotion in the Gombe National Park, East Africa, *Primates*, **18** (4), 913–21.

8   Sinclair, A. R. E. and Leakey, Mary (1986). Migration and hominid bipedalism. Letter in *Nature*, **324**, 307.

9   Laporte, L. F. and Zihlman, A. (1983). Plates, climates and hominid evolution, *South African Journal of Science*, **79**, 96–110.

10   Wrangham, Richard and Peterson, Dale (1997). *Demonic Males: Apes and the origins of human violence*, p. 59, London: Bloomsbury.

11   Wheeler, P. (1984). The evolution of bipedality and loss

of functional body hair in hominids, *J. Hum. Evol.*, **13** (1), 91–8.

12 Newman, R. W. (1970). *Op. cit.*

13 Wheeler, P. E. (1991). The influence of bipedalism on the energy and water budgets of early hominids, *J. Hum. Evol.*, **21**, 117–36.

14 Carrier, David (1984). The energetic paradox of human running and hominid evolution, *Current Anthropology*, **25** (4), 483–95.

15 Jablonski, N. G. and Chaplin, G. (1993). Origin of habitual terrestrial bipedalism in the ancestor of the Hominidae, *J. Hum. Evol.*, **24** (4), 259–80.

16 Jolly, C. J. (1970). The seed-eaters: a new model of hominid differentiation based on a baboon analogy, *Man*, **5**, 1–26.

17 Wrangham, R. W. (1980). Bipedal locomotion as a feeding adaptation in gelada baboons, and its implications for hominid evolution, *J. Hum. Evol.*, **9**, 329–31.

18 Hunt, K. D. (1994). The evolution of human bipedality: ecology and functional morphology, *J. Hum. Evol.*, **26** (3), 183–203.

19 Wood, Bernard (1993). Four legs good, two legs better, *Nature*, **63**, 587–8.

CHAPTER 6

1 Hardy, Alister (1960). Was man more aquatic in the past?, *New Scientist*, **7**, 642–5.

2 Johanson, D. C., Taieb, Maurice and Coppens, Yves (1982). Pliocene hominids from the Hadar formation, Ethiopia (1973–1977): statigraphic, chronologic and paleoenvironmental contexts, with notes on hominid morphology and systematics, *Am. J. Phys. Anthrop.*, **57**, 373–402.

3 *Tidal Forest: Siarau*. In Channel 4 documentary series 'Fragile Earth' (1983). Partridge Productions, Ltd. (The proboscis drawings are based on sequences from this film and the following one.)

4   *Long Nose, Long Tail: Borneo's Proboscis Monkey*. Film. Nippon Hoso Kyoki. Japan Broadcasting Company (photographer: Isao Takeuchi).

5   Bauer, H. R. (1977). Chimpanzee bipedal locomotion in the Gombe National Park, East Africa, *Primates*, **18** (4), 913–21.

6   Zihlman, Adrienne, Cronin, John E., Cramer, Douglas L. and Sarich, Vincent M. (1978). Pygmy chimpanzees as a possible prototype for the common ancestor of humans, chimpanzees and gorillas, *Nature*, **275** (5682), 744–6.

7   Susman, R. L. (1987). Pygmy chimpanzees and common chimpanzees: models for the behavioural ecology of the earliest hominids. In W. G. Kingsley (ed.), *The Evolution of Human Behaviour: Primate Models*, Albany: State University of New York Press.

8   White, Frances (1992). Eros of the Apes, *BBC Wildlife*, **10** (8), 40–7.

9   De Waal, Frans (1989). *Peacemaking among Primates*, Cambridge, Mass: Harvard University Press.

10 Raeburn, Paul (1983). An uncommon chimp, *Science*, **83**, 40–8.

11 Chadwick, Douglas H. (1995). Ndoki—last place on earth, *National Geographic*, **188** (1), 2–43.

CHAPTER 7

1   Darwin, Charles (1977). *The Origin of Species* and *The Descent of Man*, p. 903, The Modern Library, New York: Random House.

2   Jones, S. (ed.) (1992). *The Cambridge Encyclopedia of Human Evolution*, Cambridge University Press.

3   Wood Jones, F. (1929). *Man's Place Among the Mammals*, p. 296, London: Edward Arnold.

4   Newman, R. W. (1970). Why man is such a sweaty and thirsty naked animal: a speculative review, *Human Biology*, **42**, 12–27.

5   Schwartz, Gary G. and Rosenblum, L. A. (1981). Allo-

metry of hair density and the evolution of human hairlessness, *Am. J. Phys. Anthrop.*, **55**, 9–12.

6 Schultz, Adolph A. (1931). The density of hair in primates, *Human Biology*, **3** (3), 303–17.

7 Montagna, William and Yun, J. S. (1963). The skin of the chimpanzee (*Pan satyrus*), *Amer. J. Phys. Anthropol.*, **21**, 189–204.

8 Wheeler, P. E. (1984). The evolution of bipedality and loss of functional body hair in hominids, *J. Hum. Evol.*, **13** (1), 91–8.

9 Newman, R. W. (1970). *Op. cit.*

10 Montagna, W. (1972). The skin of non-human primates, *Am. Zoologist*, **12**, 109–24.

CHAPTER 8

1 Darwin, Charles (1977). *The Origin of Species* and *The Descent of Man*, p. 438, The Modern Library, New York: Random House.

2 Morris, Desmond (1967). *The Naked Ape*, p. 16, London: Jonathan Cape.

3 Scholander, P. F., Walters, V., Hock, R. and Irving, L. (1950). Body insulation of some Arctic and tropical mammals and birds, *Biol. Bull.*, **99**, 225–36.

4 Sokolov, W. (1962). Adaptations of the mammalian skin to the aquatic mode of life, *Nature*, **195**, 464–6.

5 Wallace, A. R. (1962). *The Malay Archipelago*, p. 301, New York: Dover Publications.

6 Elephants swam in two by two, *New Scientist*, 20 June, 1996.

7 Fischer, M. S. and Tassy, P. (1993). The interrelation between proboscidea, sirenia, hyracoidea and mesaxonia: the morphological evidence. In F. S. Szalay *et al.* (eds.), *Mammal Phylogeny*, vol. 2: Placentals, pp. 217–34, Berlin: Springer Verlag.

8 Janis, C. M. (1988). New ideas in ungulate phylogeny and evolution, *Tree*, **3** (11), 291–7.

9   Sarich, V. M. (1993). In *Mammal Phylogeny, op. cit.,* p. 113.

10   Montagna, W. (1985). The evolution of human skin, *J. Hum. Evol.,* **14**, 13–22.

CHAPTER 9

1   Montagna, W. (1985). The evolution of human skin, *J. Hum. Evol.,* **14**, 13–22.

2   Schultz, A. H. (1969). *The Life of Primates,* pp. 114 and 152, London: Weidenfeld and Nicolson.

3   Frisch, Rose E. (1984). Body fat, puberty and fertility, *Biol. Rev.,* **59**, 161–88.

4   Wheeler, P. E. (1984). The evolution of bipedality and loss of functional body hair in hominids, *J. Hum. Evol.,* **13** (1), 91–8.

5   Sokolov, V. E. (1982). *Mammal Skin,* p. 591, Berkeley: University of California Press.

6   Hillaby, J. (1962). Harvesting the hippo, *New Scientist,* **291**, 588–90.

7   Walker's *Mammals of the Sea,* Ronald Nowak (ed.). 5th edition, vol. 2, p. 1347.

8   Scholander, P. F., Walters, V., Hock, R. and Irving, L. (1950). Body insulation of some Arctic and tropical mammals and birds, *Biol. Bull.,* **99**, 225–36.

9   Sokolov, W. (1962). Adaptations of the mammalian skin to the aquatic mode of life, *Nature,* **195**, 464–6.

10   Wood Jones, F. (1929). *Man's Place Among the Mammals,* p. 309, London: Edward Arnold.

11   Montagna, W. (1985). *Op. cit.*

12   Pond, Caroline (1991). Adipose tissue in human evolution. In M. Roede, J. Wind, J. Patrick, V. Reynolds (eds.), *The Aquatic Ape: Fact or Fiction?* p. 209. London: Souvenir Press.

13   Pond, Caroline (1978). Morphological aspects and the ecological and mechanical consequences of fat deposition in wild vertebrates, *Ann. Rev. Ecol. Sys.,* **9**, 519–70 (p. 551).

14 Pond, Caroline (1991). *Op. cit.*

15 Pond, Caroline (1978). *Op. cit.*

16 *Ibid.*, p. 557.

17 *Water Babies* (1985). Golden Dolphin Productions, Ltd., producer, Robert Loader; director, Tristram Miall.

18 Pond, Caroline (1992). The structure and function of adipose tissue in humans, with comments on the evolutionary origin and physiological consequences of sex differences, *Coll. Antropol.*, **16** (1), 135–43.

19 Schultz, A. H. (1925). Embryological evidence of the evolution of man, *J. Washington Academy of Sciences*, **15**, 247–63.

20 Pond, Caroline (1992). *Op. cit.*

CHAPTER 10

1 Gould, S. J. (1993). *Eight Little Piggies*, p. 109, London: Jonathan Cape.

2 Morgan, Elaine (1990). *The Scars of Evolution*, pp. 80–104, London: Souvenir Press.

3 Schmidt-Nielsen, Kurt (1959). Salt glands, *Scientific American*, **200** (1), 109–16.

4 Peaker, M. and Linzell, J. L. (1975). *Salt glands in birds and reptiles*. Monographs of the Physiological Society, no. 32. Cambridge University Press.

5 Frey, William (1985). *Crying: The Mystery of Tears*, chapter 5. New York: Harper & Row.

6 Fossey, Dian (1983). *Gorillas in the Mist*, p. 116, London: Hodder & Stoughton.

7 Collins, E. Treacher. The physiology of weeping, *Brit. J. Ophthalmology*, January 1932, p. 8.

8 Frey, William (1985). *Op. cit.*, p. 144.

9 Steller, G. W. (1988). *Journal of a Voyage with Bering, 1741–1742*, p. 148, ed. O. W. Frost, trs. Margritt A. Engel and O. W. Frost, Stanford University Press.

10 Darwin, Charles (1965). *The Expression of the Emotions in Man and Animals*, p. 165, University of Chicago Press.

11 Harrison, R. J. and Tomlinson, J. D. W. (1963). Ana-

tomical and Physiological Adaptations in Diving Mammals, *Viewpoints Biol.*, **2**, 115–62.

12   Schmidt-Nielsen, K. (1959). *Op. cit.*

13   Darwin, Charles (1965). *Op. cit.*, p. 162.

14   Quoted in E. Treacher Collins (1932). *Op. cit.*, p. 4.

15   Montagu, Ashley (1959). Natural selection and the origin and evolution of weeping in man, *Science*, **130**, 1572.

16   Collins, E. Treacher (1932). *Op. cit.*, p. 5.

17   Montagna, W. (1982). The evolution of human skin. In *Advanced Views in Primate Biology*, pp. 35–42, Berlin: Springer-Verlag.

18   Verhaegen, Marc (1991). Human regulation of body temperature and water balance. In M. Roede, J. Wind, J. Patrick and V. Reynolds (eds.), *The Aquatic Ape: Fact or Fiction?* pp. 182–93. London: Souvenir Press.

19   Hiley, Peter (1976). The thermoregulatory responses of the galago, the baboon and the chimpanzee to heat stress, *J. Physiol.*, **254**, 657–71.

20   Robertshaw, D. (1985). Sweat and heat exchange in man and other animals, *J. Hum. Evol.*, **14**, 63–73.

21   Montagna, W. (1972) Skin of non-human primates, *American Zoologist*, **12**, 109–24.

22   Newman, R. W. (1970). Why man is such a sweaty and thirsty naked animal: a speculative review, *Human Biology*, **42**, 17–27.

23   Ingram, D. L. and Mount, L. E. (1957). Evaporative heat loss. In *Men and Animals in Hot Environments*, p. 55, Berlin: Springer-Verlag.

24   Newman, L. M., Miller, J. L. and Wright, H. (1976). Thermoregulatory responses of baboons to heat stress and scopolamine, *The Physiologist*, **13**, 271–2.

25   Gisolfi, I., Sato, K., Wall, P. T. and Sato, F. (1982). In vivo and in vitro characteristics of eccrine sweating in patas and rhesus monkeys, *J. Applied Physiology*, **53**, 425–31.

26   Johnson, G. S. and Elizondo, R. S. (1974). Eccrine sweat

glands in *Macaca mulatta*: physiology, histochemistry and distribution, *J. Applied Physiology*, **37**, 814–20.

27 Hiley, Peter (1976). *Op. cit.*

28 Montagna, W. (1972). *Op. cit.*, p. 122.

29 Mahoney, Sheila A. (1980). Cost of locomotion and heat balance during rest and running from 0° to 55°C in a patas monkey, *J. Applied Physiology*, **49**, 789–800.

30 Elizondo, R. S. (1988). Primate models to study eccrine sweating, *American Journal of Primatology*, **14**, 265–76.

31 Coon, S. C. (1955). Some problems of variability and natural selection in climate and culture, *American Naturalist*, **89**, 257–80.

32 Wheeler, Peter (1984). The evolution of bipedality and loss of functional body hair in hominids, *J. Hum. Evol.*, **13** (1), 91–8.

33 Newman, R. W. (1970). *Op. cit.*

34 Sokolov, V. (1982). *Mammal Skin*, p. 578. Berkeley: University of California Press.

35 Wheeler, Peter (1996). The environmental context of functional body hair in hominids, *J. Hum. Evol.*, **30**, 367–71.

36 Wheeler, Peter (1994). The thermoregulatory advantages of heat storage and shade behaviour to hominids foraging in equatorial savannah environments, *J. Hum. Evol.*, **26**, 4.

37 Collins, E. Treacher (1932). *Op. cit.*, p. 9.

38 Fleming, Alexander (1922). *Proc. Roy. Soc. (London)*, **B93**, 306.

39 Ferrari, R., Callerio, C. and Podio, G. (1959). Antiviral activity of lysozyme, *Nature*, **183**, 548.

40 Bodelier, V. M. W., van Haeringen, N. J. and Klaver, P. S. Y. (1993). Species differences in tears: comparative investigation in the chimpanzee. *Primates*, **34** (1), 77–84.

CHAPTER 11

1   Laitman, J. T. and Reidenberg, J. S. (1993). *Comparative and Developmental Anatomy of Laryngeal Position*, vol. 1, Philadelphia: J. B. Lippincott Co.

2   Negus, V. E. (1929). *The Mechanism of the Larynx*, London: Wm. Heinemann (Medical Books).

3   Schmidt-Nielsen, K., Bietz, W. L. and Taylor, C. R. (1970). Panting in dogs: unidirectional flow over evaporative surfaces, *Science*, **169**, 1102.

4   Darwin, C. Quoted in P. Lieberman (1985), On the evolution of human syntactic ability. Its pre-adaptive phases—motor control and speech, *J. Hum. Evol.*, **14**, 657–68.

5   Van Bon, M. J. H., Zielhuis, G. A., Rach, G. H. and van den Brock, P. (1989). Otitis media with effusion and habitual mouth breathing in Dutch pre-school children, *Int. J. Pediatri. Otorhinolaryngol*, **17**, 119–25.

6   Guilleminault, C., Cummiskey, J. and Dement, W. C. (1980). Sleep apnea syndrome: recent advances, *Internal Medicine*, **26**, 347–72.

7   McKenna, J. J. (1986). The role of parental breathing cues. An anthropological Perspective on the S.I.D.S., *Cross-Cultural Studies in Health and Illness*, **10** (1).

8   Crelin, Edmund (1978). Can the cause of SIDS be this simple?, *Patient Care*, **12**, 5.

9   Jonge, G. A. de and Engelberts, A. A. (1989). Cot deaths and sleeping position, *The Lancet*, **8672**, 1149–50.

10 Negus, V. E. (1949). *Comparative Anatomy and Physiology of the Larynx*, p. 199, London: Wm. Heinemann (Medical Books).

11 Negus, V. E. (1929). *Op. cit.*, p. 470.

12 Negus, V. E. (1949). *Op. cit.*, pp. 182–3.

13 Van den Berg, R. and Wind, Jan (1987). Has man's upright posture contributed to speech origin by lowering the larynx?, *The Netherlands ENT Society. Clin. Otolaryngol*, **12**.

14 Laitman, J. T. and Reidenberg, J. S. (1988). Advances

in understanding the relationship between the skull base and larynx with comments on the origin of speech, *Hum. Evol.*, **3**, 101–11.

15 Brul, E. Lloyd du (1974). Origin and evolution of the oral apparatus, *Front. Oral Physiol.*, **1**, 1–30.

16 Crinion, Roger (1997). (Pers. comm.)

17 Negus, V. E. (1929). *Op. cit.*, p. 54.

18 Wood Jones, F. (1940). The nature of the soft palate, *J. Anat.*, **74**, 147–70.

19 Geist, F. D. (1933). Nasal cavity, larynx, mouth and pharynx. In C. G. Hartman and W. L. Straus (eds.), *The Anatomy of the Rhesus Monkey*, chap. IX, pp. 189–209, New York: Hafner Publishing Co.

20 Negus, V. E. (1929). *Op. cit.*, p. 478.

21 Fay, Francis H. (1981). Walrus, odobenus rosmarus. In Sam H. Ridgway and J. Harrison (eds.), *Handbook of Marine Mammals*, vol. 1, 11.

CHAPTER 12

1 Pinker, Steven (1994). *The Language Instinct*, p. 164, London: Penguin Books.

2 Liebermann, P. (1985). On the evolution of human syntactic ability, *J. Hum. Evol.*, **14**, 657–68.

3 Goodall, Jane (1986). *The Chimpanzees of Gombe: patterns of behaviour*, Cambridge, Mass: Harvard University Press.

4 Gardner, R. A., Gardner, B. T. and Van Cantfort, T. E. (1989). *Teaching Sign Language to Chimpanzees*, Albany: State University of New York Press.

5 Liebermann, P., Liebermann, Jeffrey T., Reiderberg, Joy J. and Gannon, Patrick J. (1992). The anatomy and perception of speech: essential elements in analysis of human speech, *J. Hum. Evol.*, **23**, 447–67.

6 Khurana, R. K., Watabiki, S., Hebel, J. R., Joro, R. and Nelson, E. (1980). Cold face test in the assessment of trigeminal brainstem function in humans. *Ann. Neurol.*, **7**, 144–9.

7 Quoted in Elsner, R. W. (1970). Diving Mammals, *Science Journal*, **6** (4), 69–74:

8 Richet, C. (1894). La resistance des canards a l'asphyxie, *Compt. Rend. Soc. Biol.*, **1**, 244–5.

9 Elsner, R. W. (1970). *Op. cit.*, p. 72.

10 Reported in *Newsweek* magazine, 13 January, 1975.

11 Elsner, R. W. (1970). *Op. cit.*

12 Butler, P. J. and Jones, David R. (1982). The comparative physiology of diving in vertebrates, *Advances in Physiology and Biochemistry*, **8**, 244–5.

13 Hentsch, U. and Ulmer, H.-V. (1984). Trainability of underwater breath-holding time, *Int. J. Sports Med.*, **5**, 343–7.

14 Schagatay, Erika (1996). *The Human Diving Response: effects of temperature and training*, p. 96, Lund: University of Lund Press.

15 Sandon, Frank (1924). A preliminary inquiry into the density of the living body, *Biometrics*, **16**, 404–41.

16 Schagatay, Erika (1996). *Op. cit.*, p. 97.

17 Stephan, H., Frahm, H. and Baron, G. (1981). New revised data on volumes of brain structures in insectivores and primates, *Folia primatol.*, **35**, 17.

CHAPTER 13

1 Preuschoft, H. and Preuschoft, S. (1991). The aquatic ape seen from the epistemological and palaeoanthropological viewpoints. In *The Aquatic Ape: Fact or Fiction?*, p. 146, London: Souvenir Press.

2 Bronowski, J. (1974). *The Ascent of Man*, p. 401, London: BBC.

3 Morris, Desmond (1967). *The Naked Ape*, p. 74, London: Jonathan Cape.

4 Newth, D. R. (1982). One foot in sea and one on shore. Book review in spring books supplement, *Nature*.

5 Galdikas, B. M. F. (1981). Orang-utan reproduction in the wild. In Charles E. Graham (ed.), *Reproductive Biology of the Great Apes*, p. 287, London: Academic Press.

6   Morgan, Elaine (1985). *The Descent of Woman*, 2nd ed., p. 12, London: Souvenir Press.

7   Fichtelius, Karl-Erich (1991). How the aquatic adaptations of man differ from those of the gorilla and the chimpanzee. In *The Aquatic Ape: Fact or Fiction? Op. cit.*, p. 289.

8   De Waal, Frans (1989). *Peacemaking among Primates*, p. 199, Cambridge, Mass: Harvard University Press.

9   Fichtelius, Karl-Erich (1991). *Op. cit.*, p. 289.

10 Knight, C. (1991). *Blood Relations*, New Haven and London: Yale University Press.

11 Attenborough, David (1979). *Life on Earth*, p. 53, London: Collins/BBC.

12 Knight, Christopher (1991). *Op. cit.*, p. 246.

13 Montagna, W. (1982). The evolution of human skin. In *Advanced Views in Primate Biology*, pp. 35–42, Berlin: Springer-Verlag.

14 Kligman, A. L. (1963). The uses of sebum. In W. Montagna, R. A. Ellis and A. F. Silver (eds.), *Advances in the Biology of the Skin*, vol. IV, chap. 7, Oxford: Pergamon Press.

15 Kidd, Walter (1900). Notes on the hair slope in man, *Journal of Anatomy and Physiology*, **35**, 305–22.

16 Wheeler, Peter (1991). Body hair reduction and tract orientation in man: Hydrodynamics or thermoregulatory aerodynamics. In *The Aquatic Ape: Fact or Fiction? Op. cit.*

17 Evans, Peter H. Rhys (1992). The paranasal sinuses and other enigmas: an aquatic evolutionary theory, *J. Laryngol. and Otology*, **106**, 214–25.

18 Rightmire, G. P. (1990). *The Evolution of Homo Erectus*. Cambridge University Press.

19 Kern, James A. (1964). Observations on the proboscis monkey, *Nasalis Larvatus*, made in the Brunei Bay area, Borneo, *Zoologica*, **49**, 189.

20 Ellis, Derek (1986). Proboscis monkey and aquatic ape, *Sarawak Museum Journal*, no. 57 (New Series).

21 Fichtelius, Karl-Erich (1991). *Op. cit.*

22 Martin, R. D. (1983). Human brain evolution in an ecological context, *Am. Mus. Nat. Hist.*

23 Film: *A Monkey for All Seasons.* BBC: Wildlife on One. Producer: Miles Burton.

24 Crawford, Michael and Marsh, David (1989). *The Driving Force.* London: Heinemann.

25 Odent, M., McMillan, L., and Kinmel, T. (1996). Prenatal care and sea fish, *Eur. J. Obstet. Gynecol.*, **68**, 49–51.

26 Martin, R. D. (1980). Adaptation and body size in primates, *Z. Morph. Anthrop.*, **72** (2), 115–24.

27 Broadhurst, C. Leigh, Cunnane, Stephen C. and Crawford, Michael A. Rift Lake fish and shellfish provide brain-specific nutrition for early *Homo* (unpublished paper).

28 Todaro, G. J. (1980). Evidence using viral gene sequences suggesting an Asian origin of man. In *Current Arguments on Early Man*, Oxford: Pergamon Press.

29 Cronin, J. E. and Meikle, W. E. (1979). The phyletic position of *Theropithecus*: congruence among molecular, morphological and palaeontological evidence, *Systematic Zoology*, **8**, 259–69.

30 LaLumiere, Leon P. (1981). The evolution of human bipedalism: where it happened—a new hypothesis, *Phil. Trans. R. Soc. London*, **B292**, 103–7.

31 Mayr, Ernst (1963). *Animal Species and Evolution*, Cambridge, Mass: Harvard University Press.

32 Wegener, A. (1966). *The Origins of Continents and Oceans*, trs. John Biram from 4th rev. German ed., New York: Dover Publications.

33 Quoted in Sullivan, Walter (1974). *Continents in Motion: the New Earth Debate*, p. 222, New York: American Institute of Physics.

34 Sullivan, Walter (1974). *Op. cit.*

35 'Noah's Flood.' TV. A 'Third Eye' production. Transmitted by *Horizon*, BBC 2, 16 December, 1996.

36 White, T., Suwa, G. and Asfaw, B. (1994). Australopithecus, a new species of early hominid from Aramis, Ethiopia, *Nature*, **371**, 306–12.

37 Headline in *New Scientist*, 1 October, 1994.

38 WoldeGabriel, G., White, T. D., Suwa, G., Renne, P., de Heinzelin, J., Hart, W. K. and Heiken, G. (1994). Ecological and temporal placement of early Pliocene hominids at Aramis, Ethiopia, *Nature*, **371**, 330–2.

39 *New York Times*, 22 September, 1994.

Index

Page numbers in *italics* refer to information that is contained wholly or mainly in an illustration or table.

199

mating position 150–1
Mayr, Ernst 181
Medawar, Sir Peter 149
meerkats 53
Meikle, W. E. 171
menstrual rhythm 152–3
monogamy in hominids 53–4
Montagna, William 75–6, 87, 93,
    111, 115, 154
Montagu, Ashley 109
Moore, Randall 83
Morris, Desmond 77
mosaic theory 18, 19, 32
    and evolution of bipedalism 50,
    55, 58, 60
    and fatness of babies 97–8
Mount, L. E. 116
mouth breathing 126, 129, 135–6

nakedness
    in hominids 71–6, 85–6, 119,
    121, 122
    in other mammals 77–85
Napier, John 35, 40, 44–5, 54
Negus, V. E.
    on absence of speech in apes
    137
    on descended larynx 123,
    129–30, 131–2, 135
    on functions of nose 161
Newman, Russell W.
    on thermoregulation 56, 57,
    74–5, 115, 118
    on timing of bipedalism
    emergence 54, 72–3, 119, 121
nictating (nictitating) membrane
    120
nose 160–6

Odent, Michel 168
orang-utan
    locomotion 38, 41
    mating position 150–1

pachyderms 79–85, 86
palate 130, 132–5
panniculus carnosus 93
panting 112–13, 122, 125
Papio/Theropithecus split 170–1
paroxysmal atrial tachycardia
    (PAT) 141–2
patas monkey 116–17, 118
penis size 151
Peterson, D. 56
Pfeiffer, John 15, 44
philtrum 163
pigs *80*
    hair 84
    position of larynx 132
    subcutaneous fat 92
    swimming behaviour 83
Pinker, Steven 137
Pitman, Walter 173–4
plesiomorphic characters 77
Pond, Caroline 93–5, 96, 97, 99,
    100–1
proboscis monkey
    nose 161–2
    wading behaviour 64–6
pygmy chimpanzee *see* bonobo

Reidenberg, J. S. 123
reptiles
    breathing passages 124–5
    locomotion in 35–6
    nictating membrane 120
rhesus monkey, sweating 116
rhinoceros *81*, 82, 85
Richet, C. 140
Rift Valley 172
    formation 18
    increasing dryness in 26, 28
Rightmire, G. P. 159
Rodman, P. S. 48–50
Rosenblum, L. A. 73
Rowntree, V. J. 48, 50
Ryan, Bill 173–4